# U.S. Marine Corps Cold Weather Medicine Survival Course, Training and Skills



## Department of Defense

Pentagon
Publishing

# UNITED STATES MARINE CORPS

Mountain Warfare Training Center
Bridgeport, California 93517-5001

## COLD WEATHER MEDICINE COURSE

## TABLE OF CONTENTS

**UNITED STATES MARINE CORPS**
Mountain Warfare Training Center
Bridgeport, California 93517-5001

FMST.07.18
10/22/01

**STUDENT HANDOUT**

**MOUNTAIN SAFETY (WINTER)**

**TERMINAL LEARNING OBJECTIVE.** Given a unit in a wilderness environment and necessary equipment and supplies, apply the principles of mountain safety, to prevent death or injury per the references. (FMST.07.18)

**ENABLING LEARNING OBJECTIVE.**

1. Without the aid of references, and given the acronym "BE SAFE MARINE", list in writing the 12 principles of mountain safety, in accordance with the references. (FMST.07.18a)

**OUTLINE.**

1. **PLANNING AND PREPARATION.** (FMST.07.18a) As in any military operation, planning and preparation constitute the keys to success. The following principles will help the leader conduct a safe and efficient operation in any type of mountainous environment. We find this principle in the acronym "BE SAFE MARINE". Remember the key: Think about what each letter means and apply this in any type of environment.

**B** - Be aware of the group's ability.

**E** - Evaluate terrain and weather constantly.

**S** - Stay as a group.

**A** - Appreciate time requirements.

**F** - Find shelter before storms if required.

**E** - Eat plenty and drink lots of liquids.

**M** - Maintain proper clothing and equipment.

**A** - Ask locals about conditions.

**R** - Remember to keep calm and think.

**I** - Insist on emergency rations and kits.

**N** - Never forget accident procedures.

**E** - Energy is saved when warm and dry.

a. **BE AWARE OF THE GROUP'S ABILITY**. It is essential that the leader evaluates the individual abilities of his men and uses this as the basis for his planning. In his evaluation, the leader must include the group's overall physical conditioning, and the consideration of change in climate and how long the unit has had to acclimatize.

(1) Mental attitude of your group. Is morale high? How much tactical training has the group had in a particular type of terrain?

(2) Technical aspect of your group. Have they been on skis, snowshoes, etc.?

(3) Individual skills. At this point, you must choose who is most proficient at the individual skills that will be required for your mission, navigation techniques, security, call for fire, rope installations, track plans, bivouac site selection, skijoring, etc.

b. **EVALUATE TERRAIN AND WEATHER CONSTANTLY**.

(1) Terrain. During the planning stages of your mission, the leader must absorb as much information as possible on the surrounding terrain and key terrain features involved in your area of operation. Considerations to any obstacles must be clearly planned for. Will you need such things as fixed ropes, rope bridges, climbing gear, etc?

(a) Stress careful movement in particularly dangerous areas, such as loose rock and steep terrain.

(b) Always know your position. Knowing where you are on your planned route is important.

(2) Weather. Mountain weather can be severe and variable. Drastic weather changes can occur in the space of a few hours with the onset of violent storms, reduced visibility, and extreme changes. In addition to obtaining current weather data, the leader must plan for the unexpected "worst case". During an operation he must diagnose weather signs continually to be able to foresee possible weather changes.

(a) Constantly evaluate the conditions. Under certain conditions it may be advisable to reevaluate your capabilities. Pushing ahead with a closed mind could spell disaster for the mission and the unit.

c. **STAY AS A GROUP**. Individuals acting on their own are at a great disadvantage in this environment.

(1) Give the unit adequate rest halts based upon the terrain and elevation, physical abilities of the unit, combat load and mission requirements.

(2) Remember to use the buddy system in your group.

(3) Maintain a steady pace so that it will allow accomplishment of the mission when all members of the unit reach the objective area.

d. **APPRECIATE TIME REQUIREMENTS**. Efficient use of available time is vital. The leader must make an accurate estimate of the time required for his operation based on terrain, weather, unit size, abilities, and on the enemy situation. This estimate must take into account the possibility of unexpected emergencies and allow sufficient leeway to make unplanned bivouacs in severe conditions.

(1) Time-Distance Formula (TDF). This formula is designed to be a guideline and should not be considered as the exact amount of time required for your movement. Furthermore, this formula is for use in ideal conditions:

3 kmph + 1 hour for every 300 meters ascent and/or + 1 hour for every 800 meters descent.

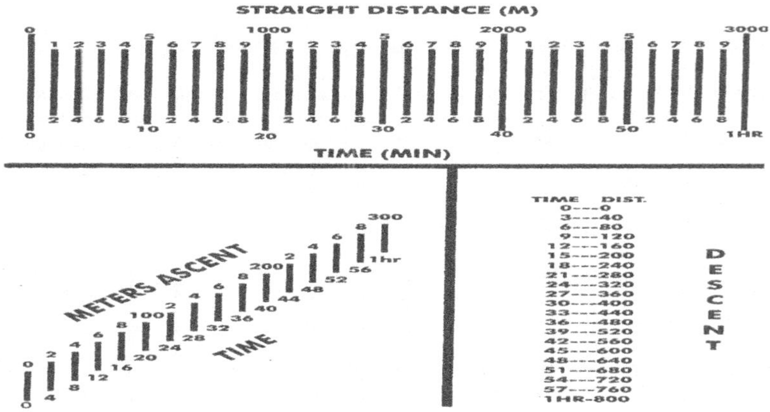


**NOTE:** The TDF is made for troops on foot in the summertime or troops on skis in the wintertime. If on foot in deep snow, multiply the total time by 2.0.

(2) Route Planning. Route cards are not to be used in place of an overlay, but as a tool to be used in route planning. Overlays/Route cards should contain the following information at the minimum:

Unit Designation:

Unit Commander.
Number of personnel.
Inclusive dates and times of movement.
Grid coordinates of each checkpoint and bivouac.
Map references.
Azimuth and distances for each leg.
Elevation gain/loss per leg.
Description of the ground.
ETA and ETD.

**ROUTE CARD**

| UNIT I.D. | | UNIT COMMANDER | | NUMBER OF PERSONNEL | | DATE AND TIME | MAP REFERENCE |
|---|---|---|---|---|---|---|---|
| | | | | | | | |
| LEG | AZM | DIST | GRID | ETA | ETD | ELEVATION | DESCRIPTION |
| | | | | | | | |
| | | | | | | | |
| | | | | | | | |
| | | | | | | | |
| | | | | | | | |
| | | | | | | | |
| | | | | | | | |
| | | | | | | | |
| | | | | | | | |

(3) As in any military operation, route planning and execution are of vital importance. Prior to departure, the unit commander must submit a route card or patrol overlay to his higher headquarters and keep a duplicate copy for himself. This preplanned route should be followed as closely as possible, taking into account changes based on the tactical situation. In non-tactical situations, the preplanned route should be followed to reduce search and rescue time in an emergency situation.

e. **FIND SHELTER BEFORE STORMS IF REQUIRED**. Under certain conditions, inclement weather can provide tactical advantages to the thinking unit commander, but by the same token it can reduce the efficiency of a unit to nil if an incorrect evaluation of the situation is made.

(1) Bivouac. If the group decision is to bivouac, then it's vital that we know the principles for an unplanned bivouac.

(a) Unplanned bivouac. The principles and techniques discussed here apply both to unplanned and tolerated bivouacs. In any survival situation, especially in a mountainous environment, the most immediate danger is from exposure to the elements. Being lost will not directly kill an individual. Starvation takes time, but hypothermia can manifest itself in a matter of hours resulting in death. Adhering to the following principles will give an individual the best chance to spend a relatively safe bivouac with the prospect of continued effort toward mission accomplishment.

1. Make shelter. The requirements for expedient shelters and the building procedures will be covered in another section. The basic requirement for protection from the elements is essential.

2. Keep warm. The retention of body heat is of vital importance; any action in which body heat is lost should be avoided. The following points should be considered:

a. Adequate shelter.

b. Insulation from the ground using branches, a rucksack, etc.

c. Wear extra clothing.

d. Use extra equipment for insulation.

e. Produce external heat while trying to conserve fuel for future use.

3. Keep dry. Being wet causes the loss of body heat 24 times faster than when dry. Adequate protection from the elements is of prime importance to prevent the onset of hypothermia.

f. **EAT PROPERLY AND DRINK PLENTY OF FLUIDS**.

(1) Food. The human body can be compared to a furnace, which runs on food to produce energy (warmth). By planning the consumption of food to suit the specific situation, adequate nutrition and extra warmth can be supplied.

(2) Water. The intake of adequate amounts of water will maintain the body in proper working order. Danger from dehydration is as high in mountain regions as in hot dry areas. Loss of liquids is easily seen and felt in hot climates; whereas in the mountains, the loss of body fluids is much less noticeable. High water intake, at least 6 quarts per day when in bivouac, 8 quarts per day when active, will help to prevent dehydration.

g. **MAINTAIN PROPER CLOTHING AND EQUIPMENT**.

(1) Clothing

(a) Our clothing has to perform an important function in our mission; therefore, when choosing our clothing we have to take into account some essential requirements.

1. Protection against wind and rain.

2. Layered and easily adjustable.

3. Lightweight and durable.

(b) To help us remember how to maintain and wear our clothing, we use the acronym, "COLD".

C - Keep clothing Clean.

O - Avoid Overheating.

L - Wear clothing Loose and in Layers.

D - Keep clothing Dry.

NOTE: In the mountains a man should never be separated from his gear. Here are some basic and essential items that should be considered during your planning stage.

(2) Required equipment:

(a) Each person in a unit should always carry an assault pack with one Marine carrying a combat load.

(b) Map and compass. Every individual in a leadership position and his assistant should carry a map and compass. The maps should be weatherproofed and extra maps should be distributed throughout the unit.

(c) Repair kit. This kit should include those items necessary to do emergency repairs on your equipment.

(d) Survival Kit. Always carried on your person. The contents of a survival kit will be covered in another period of instruction dealing specifically with survival kits.

h. **ASK LOCALS ABOUT CONDITIONS**. An often-overlooked source of information is the indigenous population of an area. Local weather patterns, rock slide areas, watering points, and normal routes can all be obtained by careful questioning. The leader must obtain current information of the actual conditions along his intended route. Of particular importance are recent precipitation and enemy sightings.

i. **<u>REMEMBER TO KEEP CALM AND THINK</u>**.

(1) Emergency situation. Having recognized that you are lost and that an emergency situation exists, the following principles should be followed:

(a) Keep calm and do not panic. At this point you must make every effort to conserve body heat and energy.

(b) Think. When an individual is cold, tired, hungry or frightened he must force himself to organize his thoughts into a logical sequence.

(c) The group must try to help itself by either finding the way back to safety or by preparing shelters and procuring food.

(d) Above all else, the group must act as a tight-knit unit. In emergency situations, individual dissension can cause a total loss of control and unit strength.

(2) If the decision is reached that the group should seek its way back to safety, several possibilities exist. In most situations, the safest approach will be to retrace the route to the last known point and continue from there. The other course of action is to get a group consensus on the present location and send out a small search party to locate a known point. This party must ensure that they mark their trail adequately to return to the group. If all attempts at finding a way back to known terrain fail, a definite survival situation exists and actions discussed later in this section must be instituted.

j. **<u>INSIST ON EMERGENCY RATIONS AND KITS</u>**. Recalling essential information from the WINTER WARFIGHTING LOAD REQUIREMENTS class, survival rations and a survival kit should always be carried.

k. **<u>NEVER FORGET ACCIDENT/EMERGENCY PROCEDURES</u>**.

(1) <u>Causes of accidents</u>. The general procedures used to handle accidents differ little in this environment, but several distinct points should be kept in mind. The most frequent causes of accidents are as follows:

(a) Overestimation of physical and technical abilities.

(b) Carelessness.

(c) General lack of observation of one's surroundings.

(d) Lack of knowledge and experience by leaders.

(e) The failure to act as a group.

(f) Underestimation of time requirements to move through mountainous terrain and underestimation of the terrain itself.

(2) Preventive measures. The only truly effective preventive measures for the above lie in the education and experience of leaders at all levels. Too often, leaders sit by watching during training and as a result have no concept of the requirements involved in the mountainous environment. Only by active involvement can a leader gain the knowledge and experience needed to effectively lead in this environment.

(3) General procedures for handling an accident. These require only a good dose of common sense as outlined below.

(a) Perform basic first aid.

(b) Protect the patient from the elements to include insulation on top and bottom.

(c) Evacuate if necessary.

(d) Send for help if required, if possible, never send a man for help alone.

(e) Send the following information regarding the accident:

1. Time of accident.
2. Nature and location of accident.
3. Number injured.
4. Best approach route to accident scene.

(4) If one man of a two-man team is injured, the injured man must be given all available aid prior to going for help. If the injured man is unconscious, he should be placed in all available clothing and sleeping gear and anchored if on steep terrain. A note explaining the circumstances, and reassuring him, should be left in a conspicuous spot. This note must also contain the following information:

(a) When you expect to return.

(b) Where you went.

(c) What you did before you left (medication, etc.).

(5) International distress signal (whistle):

(a) Six short blasts in 1 minute from person requesting help.

(b) The return signal is three blasts in 1 minute from the respondent.

(6) Other methods if help is required:

(a) Red pyrotechnics.

(b) SOS, (... --- . . .).

(c) “Mayday" by voice communications.

l. **ENERGY IS SAVED WHEN WARM AND DRY**. With the previous 11 principles in mind this one should fall right into place. Save your heat and energy by following these steps:

(1) Dress properly.

(2) Eat properly.

(3) Drink properly.

(4) Ensure shelter meets criteria.

(5) Produce external heat (fires, stove, extra clothing, etc.) to save body heat and energy for future use.

(6) Don't lose body heat by getting wet.

**UNITED STATES MARINE CORPS**
Mountain Warfare Training Center
Bridgeport, California 93517-5001

FMST.07.20
10/22/01

## STUDENT HANDOUT

## SURVIVAL KIT

**TERMINAL LEARNING OBJECTIVE**. In a winter mountainous environment, assemble a survival kit, in accordance with the references. (FMST.07.20)

**ENABLING LEARNING OBJECTIVES**.

(1) Without the aid of references, select from a given list the criteria for a survival kit, in accordance with the references. (FMST.07.20a)

(2) Without the aid of references, select from a given list one example of each criteria for a survival kit, in accordance with the references. (CWM.6.20b)

**OUTLINE.**

1. **COMPONENTS FOR A SURVIVAL KIT.**

a. The environment is the key to the types of items you will need in your survival kit. How much equipment you put in your kit depends on how you will carry the kit. A kit on your body will have to be much smaller than one carried in a vehicle.

b. Always layer your survival kit, keeping the most important items on your body.

c. In preparing your survival kit, select items that can be used for more than one purpose.

d. Your survival kit does not need to be elaborate. You only need functional items that will meet your needs and a case to hold them. The case might be a first aid case, an ammunition pouch, or another suitable case. This case should be:

(1) Water repellent or waterproof.

(2) Easy to carry or attach to your body.



(3) Suitable to accept various sized items.

(4) Durable.

e. When constructing a survival kit, you should have the following components: (FMST.07.20a)

(1) Fire starting items.

(2) Water procurement items.

(3) Food procurement items.

(4) Signaling items.

(5) First aid items.

(6) Shelter items.

2. **<u>ITEMS CONTAINED WITHIN EACH COMPONENT</u>** (FMST.07.20b)

a. <u>Fire Starting Items</u>.

(1) Matches.

(2) Magnifying glass.

(3) Flint and Steel.

(4) Lighter.

(5) Potassium Permanganate, with a container of sugar or anti-freeze.

(6) Prepackaged Tinder.

(a) Commercially Manufactured.

(b) Cotton Balls and Petroleum Jelly.

<u>b. Water Procurement Items</u>

(1) Water Disinfecting Chemicals.

(a) Iodine Tablets.

(b) Betadine Solution.

(c) Iodine Solution.

(2) Metal Container. (Serves for boiling water)

(a) Canteen Cup.

(b) Survival Kit Container.

(c) Any suitable can that contained no petroleum products.

(3) Water Carrying Items.

(a) Canteen.

(b) Plastic Bag.

(c) Plastic/Metal/Glass Container, which contained no petroleum products.

c. Food Procurement Items.

(1) Fish.

(a) Various sized hooks.

(b) Various sized sinkers/weights.

(c) Metal leaders and swivels.

(d) Small weighted jigs.

(e) Fishing line.

(i) Think about the size of fish for that environment when selecting weights and sizes.

(2) Game.

(a) Snares.

1) Commercially Manufactured.

2) Aircraft Cable.

3) Tie Wire.

(b) Bait.

1) MRE Cheese Spread or Peanut Butter Package.

(c) 550 Cord for Gill Net and Trap Construction.

1) Engineer/Marking Tape.

2) Sling shot rubber and pouch.

d. Signaling Items.

(1) Day.

(a) Mirror.

(b) Whistle.

(c) Pyrotechnics (Smoke, Pen Flares).

(d) Air Panels.

(2) Night.

(a) Pyrotechnics (Pen Flares, Star Clusters).

(b) Lights (Flashlight, Strobe, Chemlight).

(c) Whistle.

e. Shelter Items.

(1) Cordage.

(a) 550 Cord.

(b) Wire.

(c) Communication wire.

(d) Tie wire.

(2) Finger Saw.

(3) Sewing Kit with Needles for construction/repair of clothing.

(4) Teenage.

(a) Poncho.

(b) Tarp.

(c) Space blanket.

(d) Plastic trash bags.

f. First Aid Items.

(1) Band-Aids.

(a) Steristrips.

(b) Adhesive Tape.

(c) Non-stick pads, 4x4's, Gauze, Battle Dressings.

(d) Muslin Bandage.

(2) Ointments.

(a) Burn.

(b) Anti-septic.

(3) Miscellaneous.

(a) Salt.

(b) Sugar.

(c) Eye Wash.

(d) Alcohol prep pads.

(e) Suture Kit.

(f) Scalpel.

(g) Vile of Yarrow.

g. Miscellaneous items.

(1) Fingernail clippers.

(2) Compass.

(3) Notebook with pen or pencil.

(4) Wood eye screws and nails.

(5) Surgical tubing.

**NOTE**: It is assumed that the Marine is always carrying a high quality fixed bladed knife, a multi-tool knife, and a sharpening stone.

**UNITED STATES MARINE CORPS**
Mountain Warfare Training Center
Bridgeport, California 93517-5001

LP
10/22/01

**STUDENT HANDOUT**

**COLD WEATHER CLOTHING**

**PURPOSE**. The purpose of this period of instruction is to familiarize the students with the cold weather clothing systems used in the Marine Corps. Also, we will discuss cold weather dressing theory, layers, fabrics, and the individual components of the ECWCS. This lesson relates to survival.

**OUTLINE**.

1. **COLD WEATHER ENVIRONMENTS**. Cold weather can be characterized as two types: wet cold and dry cold.

   a. Wet Cold. Wet cold conditions occur where temperatures are near freezing above 14°F. Variations in the day and night temperatures cause alternate freezing and thawing. These conditions are often accompanied by wet snow and rain causing the ground to become mushy and muddy. With these conditions, Marines require clothing that consists of a waterproof or water repellent, wind resistant outer layer, and an inner layer with sufficient insulation to provide protection in moderately cold weather of 14°F or above.

   b. Dry Cold. Dry cold conditions occur when average temperatures are lower than 14°F. The ground is usually frozen and the snow is dry. These low temperatures, plus wind, increase the need for protection of the entire body. For these conditions, Marines require clothing that will provide insulation for the body for a wind-chill factor of -80°F. A water repellent, wind resistant outer layer must protect the inner layers of insulation.

2. **COLD**. To help us remember some of the basic principles of wearing our clothing, we use the acronym "COLD".

   **C** - Keep clothing clean.

   **O** - Avoid overheating.

**L** - Wear clothing loose and in layers.

**D** - Keep clothing dry.

Here are some theories behind the acronym "COLD":

a. **C** - Keep Clothing **CLEAN**. Clothing keeps you warm by trapping warm air against your body and in the pores of the clothing itself. However, if these pores are filled with dirt, sweat, or other grime, it will not be able to do its job as efficiently. Therefore, your clothes should be kept as clean as possible to keep you as warm as possible.

b. **O** - Avoid **OVERHEATING**. Everyone naturally assumes that the more clothes you have on, the warmer you will be. This is true up to a point, and that point is when your body starts overheating and sweating. A Marine, who is on a forced march, or digging a position, will be warmer through body activity than one who is just standing guard. However, if they are both dressed the same, the one that is active will start to overheat, getting his clothes wet and dirty, and his body too hot. The key to surviving under this condition is not to be hot and snug, but comfortably cool, not cold, but cool. If at any time you are sweating, you are too hot. Sweating is a sign that your body wants, and needs, to cool down. Let the environment cool you down, not your body sweat. This may be as simple as opening a button or two, instead of removing a whole layer of clothing. Once you stop your work, or feel yourself getting cold, bundle up again just enough to keep cool. Allowing just enough clothes and body activity to keep you appropriately warm, and the environment to cool you down, will keep your clothes from getting sweaty and dirty, and therefore more effective. In addition to over heating and causing problems with your clothes it can also cause problems with your body. Overheating such as dehydration, heat exhaustion, and hypothermia can cause several cold weather injuries.

c. **L** - **LOOSE and LAYERED**

(1) Loose Clothing. You want to keep your clothes loose for comfort. If clothing is too tight:

(a) It may act somewhat like a tourniquet, causing blood in your extremities, i.e., arms, legs, fingers, toes, etc., to pool there, not allowing it to get back into your body core and rewarm, thereby causing that limb to get cold.

(b) Little or no air can be trapped between your body and clothes. It's this warm air that keeps you warm, not the clothes. How this works will be discussed under the next topic.

(2) Layering. This is another important principle for staying warm in the cold. You can compare this to your house. Your house has several layers, not just one, to keep you warm. It has shingles and a roof, a wood frame, siding and insulation, walls, foundation and floor; and to heat it you have a furnace. The furnace heats the air and keeps you warm. The layers are what hold it around you. The first wall holds a warm cushion of air around you, maybe only letting 25% of the warm air escape. However, the second wall or insulation will capture part of that, holding another warm layer of air, until only a fraction of warm air escapes to the outside. Even when a strong wind hits your house blowing away one layer, you still have several others. Your body works along the same principle, with your body being the furnace and your clothing layers, the walls. The more layers, the more warm air that is going to be trapped. Strangely enough, several thin layers working together will work better than one thick layer working alone.

c. **D** - Keep Clothing **DRY**. The final letter warns you to keep your clothing dry. This means not only from the outside, such as putting on rain gear during sleet or when walking through wet snow, but also from the inside, such as taking a layer off when you start sweating. Once your clothes are wet, the water or sweat starts to evaporate, drawing off warmth with it. Always keep your clothes dry.

3. **TYPES OF LAYERING**. As discussed earlier, layering is a very important aspect to staying warm. There are, however, different types of layers to use in the cold weather clothing system:

a. Vapor Transmission Layer. Better called a "sweat transfer layer", this layer soaks up your body moisture and draws it away from your body to keep it dry. Significant progress has been made with such synthetics as polypropylene, which draws water away from the body, but stays dry.

b. Insulating Layer. This is the layer that holds the warm layer of air around your body. Preferably, it is made of polyester pile, but wool is adequate.

c. Protective Layer. This not only protects the insulating layer from getting dirty, but also from getting wet. It should be made of a wind resistant/water repellent substance.

NOTE: These are the three main layers to consider. There may be times when one or more layers are not used, or when the insulating layer may be several layers thick.

4. **THE CLOTHING SYSTEMS**.

a. **GENERAL**. There are three clothing systems currently in use in the Marine Corps; the new Extended Cold Weather Clothing System (ECWCS) that is currently coming into service, the old M-1950 Cold Wet/Cold Dry seven layer system and the Specialty Clothing System developed to fit the specific needs of aviation and maintenance personnel.

(1) Extended Cold Weather Clothing System (ECWCS). The ECWCS has recently been approved for Marine Corps use. ECWCS was developed to provide a lighter weight, less bulky clothing system that was better suited to the modern cold weather battlefield. This system uses recently developed synthetic materials to provide warmth and handle moisture much better than the older standard clothing system. The ECWCS is a layered insulating system adjustable to personal preference, metabolism, and prevailing weather conditions. It is designed to maintain adequate environmental protection between +40°F and -25°F (4°C and -13°C), with the Extreme Cold Weather Boots protecting down to -50°F (-45°C). This system uses moisture management principles to transfer perspiration away from the skin so that the user will remain warm and dry. In cold, wet, and arctic environments, it is recommended that Marines use only clothing items in the ECWCS system. It is especially recommended that Marines not combine use with any items that are not made with wool or wool blends. This system is not cumbersome in weight and bulk, with the total system weighing approximately 18 pounds. It is easy to maintain in both field and garrison environments; however, because of the unique characteristics of the state of the art material the special use and care instructions must be carefully followed.

(2) M-1950 Cold Weather Uniform. The standard cold weather clothing system was developed during and after the Korean War and has been in use with very little change ever since. It consists of uniforms for cold wet and cold dry environments and is made up of 5 to 7 layers. This clothing system is presently being phased out of the USMC inventory and is being replaced by the ECWCS.

(3) Specialty Uniforms. The standard cold weather uniform system and the ECWCS may not meet the requirements of all aviation and maintenance personnel. As result a specialty uniform system has been developed for their use. Generally, this specialty uniform system utilized items of cold weather clothing found in the supply system. These items of equipment are not stored or obtained from the Training Allowance Pool (TAP). Specialty uniforms are not issued by MCMWTC and will not be discussed any further.

b. **ECWCS CLOTHING ITEMS**.

(1) General. The Marine Corps has recently completed a research and development project, which resulted in the adoption of a new, cold weather, state of the art clothing system. Described below are the component items of that system which combines many new items with a few items that have remained from the older M-1950 clothing system.

(2) Long Underwear. The cold weather underwear consists of the undershirt and drawers (Undershirt, Cold Weather, Polypropylene/capilene, and drawers, Cold Weather, Polypropylene/capilene).

(a) Description of Item. The polypropylene undershirt is a buff-colored turtleneck that has a center front zipper, which extends to the middle of the chest area. The polypropylene drawers, also buff-colored, serve as a base layer to protect the lower extremities. The capilene undershirt and capilene drawers are similar except they are lightweight and olive-green in color.

(b) Concept of Use. The underwear layer next to the skin acts as a moisture wicking layer and serves to draw moisture away from the skin, transferring it to the outer layers of the system. The wearing or use of issue cotton undershirt and underwear will negate the wicking action of the polypropylene underwear. The cotton fibers will hold moisture next to the skin. Do not wear cotton undershirts and underwear when using this clothing.

(3) Cold Weather Shirt. The cold weather shirt is a new item similar in design to the OD sleep shirt (Shirt, Cold Weather, Polyester Fiber pile).

(a) Description of Item. The shirt is brown-colored, has reinforced shoulder and elbow patches, a convertible turtleneck collar, front zipper, elastic draw cord waist, hook and pile cuff tabs, two chest cargo style pockets, and two lower hand-warmer pockets.

(b) Concept of Use. The polyester fiber pile shirt serves as the primary insulating layer on the upper body.

(4) Bib Overall. The bib overall (Overall, Bib, Cold Weather, Fiber pile) is a new item which Marines will seldom need unless deployed to operational areas where temperatures are in the extreme cold range.

(a) Description of Item. The overall is brown-colored, has elastic suspenders with quick release buckles, and full-length zippers at the side seams.

(b) Concept of Use. The bib overall is used as an additional layer for temperatures below -25°F (-32°C).

(5) ECWCS Cold Weather Liner, Trousers/Coat. The full closure liners serve as an insulating layer. The trouser liner and coat liner are designed to be worn as an added layer of insulation during periods when Marines are stationary, i.e., standing sentry duties, or during periods of intense cold.

(a) Description of Item. The standard coat and trouser liners can be worn independently of their respective outer garment.

(b) Concept of Use. The coat and trouser liners serve as an insulating layer. The trouser liner and coat liner are designed to be worn as an added layer of insulation during periods when Marines are stationary, i.e., standing sentry duties, or during periods of intense cold.

(6) Trousers, Cold Weather Field, Nylon and Cotton. The olive green or four color camouflage printed cold weather field trousers (Trousers, Cold Weather, Field, Nylon and Cotton) are standard items of cold weather issue and are also used in the ECWCS system.

(a) Description of Item. Characteristics of the field trousers are side hanging pockets, hip pockets, cargo pockets, draw cords at the trouser bottoms and adjustable waist straps.

(b) Concept of Use. The field trousers serve as a durable insulating layer to be worn over the liners when the outer extended cold weather trouser is not needed.

(7) ECWCS Camouflage Parka. The extended cold weather camouflage parka (Parka, Extended Cold Weather, Camouflage) is a new item.

(a) Description of Item. The parka has an integral hood, two breast inside map pockets, which can be opened without unzipping the parka, two large lower cargo pockets and a two-way, full-front zipper to provide full face protection, leaving only the eyes uncovered. There is an elastic draw cord at the hem, hook and pile closures at the wrist tabs, underarm ventilation and a rank tab at center chest.

(b) Concept of Use. The parka serves as part of the windproof, waterproof layer in the system. The polytetrafluoroethylene (PTFE) laminate in the garment has the property to repel water while allowing perspiration to be expelled.

(8) ECWCS Camouflage Trousers. The extended cold weather camouflage trousers (Trousers, Extended Cold Weather, Camouflage) are a new item.

(a) Description of Item. The trousers have seat and knee patches, pass through pockets, and inserts in the seams of the leg openings to allow easy donning and doffing without removing the boots.

(b) Concept of Use. The trousers serve as part of the windproof and waterproof layer in the system. The PTFE laminate in the garment has the property to repel water while allowing perspiration to be expelled.

(9) Snow Camouflage Parka and Trousers. The snow camouflage parka and trousers (Parka, Snow Camouflage and Trousers) are standard items from the M-1950 issue.

(a) Description of Item. The hooded white parka has drawstrings for adjustment at the waist draw cord, side pockets, a hip pocket, knee pleats and drawstrings at the ankles of the trouser. The parka and trousers, snow camouflage, are to be worn for whatever type of camouflage is required in snow covered terrain.

(b) Concept of Use. The over white parka and trousers are used as a camouflage outer layer in snow covered terrain and is not a substitute for an outer garment. It is worn over the parka and trousers.

(10)Hood C/W (Balaclava). The head wear in the ECWCS system cold weather system.

(a) Description of Item. The hood consists of a wool, knitted cap, which covers the entire neck and face with holes for the eyes and nose and is a pullover ski mask style that comes in either green or black.

(b) Concept of Use. The cap is intended to provide protection in cold weather to the neck and face.

c. **HAND WEAR**.

(1) General. The standard hand wear items are: Glove inserts, gloves, mitten inserts, mitten shells (cold weather and snow camouflage) and mitten set. These items are carry-over items from the standard M-1950 cold weather issue. These items are considered part of the ECWCS issue. The gloves and glove inserts are unit supply items, while the mittens, mitten inserts and camouflage shells are Training Allowance Pool items.

(2) OD Knitted Glove. The knitted OD glove insert (Glove Insert) is the standard insert that Marines issue as organizational equipment.

(a) Description of Item. The olive drab glove inserts can be worn on either hand. All parts of the glove, except the cuff, are seamless knit. The cuff is a true rib knit.

(b) Concept of Use. The glove inserts are worn for added warmth with light duty gloves.

(3) Black Field Gloves. The black gloves (Gloves, Men's and Women's, Light Duty) are the standard field gloves normally issued to Marines as organizational equipment.

(a) Description of Item. The gloves are black, slip-on style, all leather, with a buckle strap of black tape on back for wrist closure. The gloves are either a one-piece back and front or a two-piece back design.

(b) Concept of Use. The gloves are used for light work duty. They may be worn alone or, for additional warmth in cold conditions, and may be worn with the cold weather glove inserts.

(4) OD Mitten Insert. The knitted OD mitten inserts (Mitten Inserts, Cold Weather w/Trigger Finger) are part of the Training Allowance Pool (TAP) issue.

(a) Description of Item. The mitten inserts can be worn on either hand and have separate sections for the thumb and index finger.

(b) Concept of Use. The mitten inserts are used as an insulating layer under the cold weather mitten shell.

(5) Mitten Shells, Cold Weather w/Trigger Finger. The cold weather mitten shells (Mitten shells, Cold Weather w/Trigger Finger) are part of the TAP issue.

(a) Description of Item. The color of the leather is saddle brown. The mittens are slip-on style with a trigger finger. The mitten has an elastic webbing and tape loop for a suspension cord at the cuff opening and an adjustable wrist strap on the back of the mitten shell. The back of the thumb, hand, and trigger finger down to the cuff of the mitten is lined with insulation for added protection.

(b) Concept of Use. The trigger finger mittens are to be worn with or without the wool/nylon mitten inserts and in areas too cold for leather gloves and not cold enough for arctic mittens.

(6) Extreme Cold Weather Mitten Set. The extreme cold weather mittens (Mitten Set, Extreme Cold Weather) are part of the TAP issue.

(a) Description of Item. The mitten set consists of a shell with a removable liner and a suspension harness. The palm and front of the thumb of the shell is leather. The remainder of the mitten shell is fabric. The back of the mitten shell includes a wool pile material used for warming the face. Two adjustment straps and buckles are included at the back of the mitten near the cuff. The liner is attached to the shell by snaps. The harness includes a suspension piece (cord) and a breast piece (strap).

(b) Concept of Use. The mitten set is worn over the wool mitten insert and trigger finger mitten shells to provide environmental protection to the hands in a cold weather environment. The suspension harness is worn over the camouflage parka. The cord is placed around the neck and the strap is adjusted to lie on the chest to keep the cord in place. The ends of the cord are tied to the loop on the outer shell of each mitten. The outer shell of the mitten set can be separated from the insulating liner for drying.

(7) Camouflage Mitten Shells. The camouflage mitten shells (Mitten Shells, Snow Camouflage, Cotton, White, Two Finger) are part of the TAP issue.

(a) Description of Item. The mitten shells have an adjustable strap on the back across the wrist and an elastic take-up at the hem of the mitten opening. The thumb side has an opening for the trigger finger.

(b) Concept of Use. The mitten shells are used as the hand-wear camouflage cover when worn over the mitten shells, cold weather, and mitten set, extreme cold weather, in snow environments.

(8) Polypropylene Glove Liners. The polypropylene glove liners are currently not in the system for issue, but can be purchased through the civilian market.

(a) Description of Item. The polypropylene glove liner is a thin synthetic glove, usually blue in color.

(b) Concept of Use. The polypropylene glove liner is worn on the hands under any hand wear currently used by the Marine Corps. It's ideal for extracting moisture from the hands as a vapor transmission layer and can do light work as contact gloves.

d. **ACCESSORY ITEMS**.

(1) General. The items in this section are considered part of the ECWCS issue. Some of these items are new to the Marine Corps while some are carry over items from the M-1950 issue.

(2) Suspenders. The suspenders (Suspenders, Trousers M-1950) are a carry over item and are used with the field trousers.

(a) Description of Item. The olive drab suspender straps are scissor-back style (cross over in the back). The suspenders have two slide buckles and two hooks which attach to the trousers.

(b) Concept of Use. The suspenders are to be used with the trousers, extended cold weather, camouflage.

(3) Head-over Scarf. The head-over scarf (Scarf, Head-over) is an item borrowed from our NATO allies. This item enables Marines to regulate their body temperature.

(a) Description of Item. The head-over scarf is a circular knitted wool tube 2 feet long and 9 inches wide lay flat, open at both ends, with the face of the fabric lightly brushed.

(b) Concept of Use. The head-over scarf is to be wrapped around the neck, pulled over the head and ears, or pulled down over the neck and lower back.

e. **CARE, USE, AND MAINTENANCE OF ECWCS**.

(1) General. The individual Marine is responsible for keeping his ECWCS items in good serviceable condition. This is his uniform. It will not provide optional performance unless it is kept clean, maintained in good repair, and stored properly. The ECWCS will protect him only if he/she takes care of it and wears it properly. Check the label to see if the size is correct. This is extremely important in order to achieve maximum user satisfaction using the layering principle. ECWCS IS DIFFERENT. Pay particular attention to cleaning instruction for layers 1 and 4, polypropylene underwear and parka/trousers, extended cold weather, camouflage, as these items are made of state of the art materials and require added care.

(2) Donning and Doffing Procedures. The ECWCS is an insulating system consisting of the following five primary layers (including the over whites, when necessary) and accessories:

(a) Layer 1 - Polypropylene undershirt and drawers.

(b) Layer 2 - Bib overall, cold weather shirt and trouser liner.

(c) Layer 3 - Coat liner and field trousers.

(d) Layer 4 - Extended cold weather camouflage parka and trousers.

(e) Layer 5 - Snow camouflage parka and trousers (over whites).

(3) Layering. Layers 1 and 4 are always worn. Add layers 2 and 3 as necessary to stay warm. The bib overalls in layer 2 are normally worn for temperatures below -25°F (-32°C). Remove layers 2 and 3 as necessary to avoid overheating when on the move. The polypropylene underwear has the ability to draw moisture away from the skin and transfer it to the outer layers of the system. Beginning with layer 1, add layers 2 and 3 as the temperature drops. Layer 4 is the outer layer of the ECWCS when woodland camouflage is required. Layer 5 is the outer layer of the ECWCS when snow camouflage is required. Layer 5 is the outer layer of the ECWCS when snow camouflage is required. Layer 5 is not a substitute outer garment, but is worn over layer 4 only as camouflage. Adjust layers according to preference, metabolism, and weather conditions.

(4) Inspection. Examine the ECWCS items regularly for tears, punctures, or damage to the material. Punctures on the outer layer will produce leaks and eventually ruin the material if not properly maintained. Repairs should be made as soon as possible.

(5) Attaching Rank Insignias. Attach rank insignia on the parka to the rank tab, which is provided at the center of the chest. Either the pin on or sewn on rank insignia may be used. Be careful not to puncture or snag the outer layer of the material when attaching rank as punctures will produce leaks.

(6) Cleaning. Clean ECWCS clothing items regularly when in use. Dirty clothes wear out quickly because dirt cuts textile fibers and retains moisture from perspiration. Prior to laundering and drying, make sure all the draw cords are tied together, all zippers are zipped, and all snaps and hooks are fastened. Securing these items will result in a better-laundered garment. When laundering, use delicate or gentle fabric wash cycle or by hand, using cold water (up to 85°F/29°C) and cold water laundry detergent. Rinse in clean cold water. DO NOT USE BLEACH OR STARCH. Tumble dry at the lowest fabric cycle, delicate/gentle. Do not exceed 90°F/32°C. Remove immediately at the end of drying. AVOID OVER DRYING. To drip dry, remove water and place on a rustproof hanger. **DO NOT PRESS**.

(7) Water Repellency. If the fourth layer (Parka/Trousers, Extended Cold Weather, Camouflage) of ECWCS leaks and inspection has revealed no rips or tears, wash garments in mild powdered detergent. Detergents used in cleaning affect water repellent qualities. DO NOT WASH GARMENTS IN LIQUID DETERGENT. When liquid detergents are used, they leave a chemical residue, which actually reduces the waterproof properties of the fabric. To restore the water repellency of the parka/trousers, occasionally steam garments with an iron on steam setting being careful to hold the iron about 1/2 inch above the garment. **REMEMBER DO NOT PRESS**.

f. **FOOTWEAR**.

(1) General. USMC footwear consists of two types: The old Vapor Barrier system and the new ski march boot/sock system.

(2) The Vapor Barrier (VB) boot system consists of three items: The cushion-sole wool socks, the black cold weather boot and the white extreme cold weather boot. The barrier (VB) boots. These boots use an inner and an outer boot made of rubber and filled with either wool fleece or closed cell foam (neoprene) insulation. The rubber acts to stop the movement of moisture from the feet. The moisture in the air transfers heat quickly. By trapping the moisture, the boots trap heat. The boots also act to keep the moisture out. New Socks, Men's, Nylon, Cushion Sole Stretch type, OD-106 should be utilized by Marines. Carry dry socks and change socks at least 3 times a day when wearing VB boots. When possible, the VB boots should be removed for at least a few hours a day to allow the feet to breathe and dry out.

(a) VB Boots. There are two types of VB boots:

1. Boots, Cold Weather (Type 1, Black). These boots are worn in the cold wet environment and will protect the feet down to -20°F.

2. Boots, Extreme Cold Weather (Type 2, White). These boots are worn in the cold dry environment and will protect the feet down to -50°F.

(b) Ski March Boot System. The system consists of several layers including vapor transmission socks, insulating socks, vapor barrier socks, the boot itself, and several different over boot designs.

1. Components of the Sock system: The sock system currently undergoing test in the Marine Corps is a three sock layered system.

   a. The first layer. Is a lightweight single layer polypropylene sock. It should fit snugly. It is designed to wick moisture away from the foot and prevent blisters by reducing friction. The intermediate layer of the system is a vapor barrier sock. (This layer is only worn in extreme cold temperatures. Working on the same principle as the Vapor Barrier (VB) boot, it traps all of the heat created from the feet. The problem is that it also traps all of the moisture.) This sock should be worn between the vapor transmission and the insulating layers. This keeps the foot warm and protects the insulating layer from perspiration. Never wear the VB sock over the insulating sock, as it will cause it to become saturated and lose its heat retention properties. Remember the "D" in COLD. Also, be careful of using the VB sock when it is warm as it can cause blisters due to excessive sweating. The second layer is a hook stitch pile fiber made of 50% wool and 50% polypropylene. This combination provides the warmth needed for prolonged ski movements and still allows the moisture to pass through the sock. When looking at this sock, you will see that there is a smooth side and a rough side. The smooth side is worn on the inside next to your foot.

2. Use of the Sock System. Two pairs of the vapor transmission and insulating socks are issued. This enables the wearer to continually rotate the socks, allowing the other pair to be dried using whatever method is available. Body heat works well.

3. Care of the Sock System. Care of the sock system falls into two areas, field care and garrison care.

   a. Field care. It is important to keep the socks as dry and clean as possible to prevent them from losing their specific properties. Crinkle and shake your socks to keep them as free as possible from dirt and body oils, which render them less effective. This is in keeping with the "C" in COLD.

   b. Garrison care. Wash the polypropylene inner socks in the same manner as the polypropylene long underwear. The wool socks should be washed in cold water with a mild detergent such as Woolite.

4. Ski March Boot. The ski boot now undergoing final testing is the ALICO single ski boot. It has a box-toe and a grooved heel that is designed to work

with our standard issue ski and the NATO 120 binding. The boot has a fully gusseted leather tongue and collar with Thinsulate and Vapor insulation.

a. Felt Liners. The felt liners are designed to add insulation, absorb moisture that would otherwise be absorbed by the boot itself, and to form to the foot for a more comfortable fit. The boot is provided with two liners to allow a wear/dry rotation. Sleeping with one liner next to your body will both dry and warm it prior to putting the boot on.

b. Sizing. Proper sizing of the boot is perhaps the most important thing. Done wrong, the Marine may suffer from blisters or frostbite. Sizing must be done with the felt liner in the boots and wearing all three socks of the previously discussed sock system. Normally, 1 to 1 1/2 sizes larger will be required for proper fit. The boot should fit snugly in the heel area to avoid blisters, but not so tight that it cuts off circulation. The toes should have some movement, but avoid side slippage, which not only causes blisters but also reduces control while skiing.

c. Breaking in. New boots must be broken in gradually. Wear for no more than four hours at a time for the first few days to prevent blisters as well as foot fatigue.

d. Waterproofing. There are many products on the market that will work, but here at MWTC we use snow seal. Instructions for use are on the package.

5. Care of the Ski March Boot. Caring for your ski march boot is much the same as any other leather boot. Dry your boot whenever possible, being careful not to use open flames or any method that will dry the boot too quickly. Strive to keep the boot dry to prevent freezing. Using foot powder to absorb excess moisture is okay as long as it is kept to a minimum; otherwise, the fibrous parts of the boot can become clogged and reduce effectiveness.

6. Gaiters. Also known as "Super" gaiters, these are worn with the ski boots. It is a nylon Gore-Tex legging that is fully insulated. To place on the boot, feed the toe of the boot through the front hole on the bottom of the gaiter. Slide the heel of the boot through the rear hole in the rubber bottom. Pull the toe of the gaiter over the toe of the boot, ensuring that the rubber seal is snug against the welt of the boot. If the rubber seal will not stay in place along the toe of the boot, a light coat of Purple Klister will help stick it in place.

7. Over boots. Over boots, as their name implies, are worn over the entire ski march boot system. The over boots are fully insulated and have a hard sole for walking, they are not; however, designed to be used for skiing.

a. Care of the Gaiters/Over boots. Caring for the Super gaiters and the Over boots is essentially the same. Keep as clean and dry as possible. Open the gaiter occasionally while wearing to allow condensation to evaporate. If the rubber parts start to dry out, coat them with a silicone spray.

g. **COLD WEATHER PERSONAL EQUIPMENT**.

(1) General. Personal equipment for use in cold weather environments is especially designed to provide protection and be as lightweight as possible.

(2) Sleeping System. The sleeping system consists of a sleeping bag, an insulated sleeping mat, and a waterproof bag.

(a) There are two types of sleeping bags:

1. Sleeping Bag, Type I, intermediate cold, for temperatures down to +10°F, uses polyester batting for insulation and weighs 7.5 lbs.

2. Sleeping Bag, Type II, extreme cold, for temperatures down to -50°F, uses waterfowl feathers, down, and polyester batting for insulation, weighs 9.5 lbs.

(b) Sleeping Mat. The sleeping mat replaced the old pneumatic mattress. It provides excellent insulation from ground cold and can be used for sitting, sentries, when consolidating following assaults, and in ambush positions when personnel must lie prone for long periods of time.

(c) Waterproof Bag. This is used to protect the sleeping bag from getting wet. Both bags are difficult to dry once wet and care should be taken to keep them as dry as possible.

(3) Load Carrying Equipment. Marines are now issued the LCS-88 pack for use in cold weather/mountainous environments. This pack has an internal frame, fully adjustable suspension, a map flap, three external ski tunnel pockets, an internal divider for the zipper opened sleeping bag compartment, a radio pocket, and numerous attachment points for ALICE equipment. Packing of this pack is covered in the Winter War fighting Load Requirements class.

(4) Miscellaneous Gear.

(a) Sunglasses. Snow blindness is a very real problem in snow-covered terrain. Snow blindness is painful, it requires bed rest to treat, and a snow blind Marine is a liability to the unit. Always use the issued sunglasses or sunglasses with side shields designed to filter out ultraviolet rays. Even on overcast days, the possibility exists for snow blindness to develop. Leaders must ensure that Marines have their sunglasses and require their wear.

(b) NBC Protective Mask. There is a cold weather kit for the NBC protective mask. Be sure to winterize the NBC protective mask prior to cold weather deployments.

(c) Canteens. The plastic canteen will freeze very quickly in cold weather. If used, it must be carried in the interior of the clothing or deep in the pack wrapped in spare clothing. The two-quart, collapsible canteen is very useful in cold weather operations, but it too must be carried next to the body. The one-quart, stainless steel vacuum bottle is an insulated canteen for cold weather use. It will keep water from freezing, but is difficult to use.

(d) Personal Survival Kit. A Marine separated form his gear can soon become a casualty and is not effective to his unit. Above that, even during training operations, the cold can easily chill an unprepared Marine. Each Marine should carry survival items on his person at all times in the cold weather environment. Some FMF units frequently assigned to cold weather operations have improvised most of the contents of this suggested kit (TAP issue items). A suggested list, not all inclusive, includes the following items: a small sharp knife, matches in a waterproof container, five meters of strong line, a small flashlight, sunburn preventive cream, chap stick, a candle, sunglasses, a small brush, a whistle, thermos bottle, emergency rations (1300 calories, one MRE), canteen cup, poncho, and a space blanket.

h. **TRICKS OF THE TRADE**. The final subject we will talk about is some of the things that can be done to take care of yourself, and your gear.

(1) Ideas for Taking Care of Yourself:

(a) Cold Fingers. When using gloves or even mittens, your fingers tend to get cold. Simply pull your fingers out of their compartments, keeping them in the gloves, and make a fist. Your warm palm surrounding the fingers will warm them up. You can also swing your arms in big circles to force warm blood down into your fingers.

(b) Sleeping Bag. Along with the sleeping mat, branches or MRE sleeves will help insulate the bottom. For the top, any rubberized gear will hold the heat in the bag. However, in the morning, the surface condensation should be quickly dried off. A balaclava should be worn while sleeping. Also ensure that the flap is pulled over the zipper to prevent the zipper from freezing shut.

(c) Food Before Sleeping. You heat your sleeping bag, it doesn't heat you. Therefore, just before going to sleep, eat something high in carbohydrates like noodles, cereal or crackers to give you fuel to burn during the night. A hot liquid will also do the job.

(2) Ideas for Your Clothes:

(a) Waterproofing. As clothes get old and worn, they lose their water repellency. The water repellency can be restored on the ECWCS as discussed earlier, NEVER use Scotch Guard on this system.

(b) Storing Clothes. Although you don't want to sleep with your clothes on, you also don't want to leave them out to get cold. Therefore, if they are clean, you can put them inside your sleeping bag along the sides. Boots can be placed inside an inside-out waterproof bag, and then placed over the foot of your sleeping bag.

(c) Conventional Canteens. It is preferred over the arctic canteen. Keep it 2/3 full to prevent freezing, and wrap it in your pack, or keep it next to your body. When going to sleep, they can be filled with hot water, and placed inside the bag. This will not only keep them from freezing, but will also warm the bag.

(d) Drying. For drying small items like socks or gloves, wear them next to your body during the day, and rotate as needed. Larger items can be dried in the tent. If your clothes are wet, hang them in the tent to dry, sleep in dry clothes, and wear the wet clothes, drying them out with body heat. Bringing them into your sleeping bag at night can dry out small, damp items such as gloves and socks.

## UNITED STATES MARINE CORPS

Mountain Warfare Training Center
Bridgeport, California 93517-5001

LP
10/22/01

**STUDENT HANDOUT**

**WINTER WARFIGHTING LOAD REQUIREMENTS**

**PURPOSE**. The purpose of this period of instruction is to introduce the student to the different levels of gear requirements in accordance with FMFM 7-21 Tactical Fundamentals for Cold Weather War fighting and MWTC's uniform codes. This lesson relates to all ski training and tactical evolutions.

**OUTLINE**.

1. **BASIC UNIFORM REQUIREMENTS**.

   a. The following list of uniform items comprises the basic cold weather uniform. This list may vary depending upon the severity of the weather, the activity level of the Marine, and the individual metabolism of the Marine. However, the unit leader should dictate the outer camouflage layer.

      (1) ECWCS Parka and Trousers

      (2) C/W Trousers w/Suspenders

      (3) C/W Hat or Balaclava

      (4) Polypropylene or Capilene long underwear

      (5) VB Boots or Ski March Boots w/appropriate sock system and gaiters

      (6) Gloves with Inserts or Trigger-finger Mittens

      (7) Over whites (parka, trouser, and over mittens)

      (8) Helmet w/Camo cover (white)

b. As part of the basic cold weather uniform, each man should be required to have in his possession at all times some required pocket items. These seven items should be carried in the pockets of your ECWCS uniform:

(1) Pocket knife

(2) Whistle

(3) Pressure Bandage

(4) Chap stick and sunscreen

(5) Sunglasses

(6) Survival Kit and rations (fire starting material, signaling material, food gathering material, water procuring material, and some high energy, lightweight snacks).

(7) Notebook with pen/pencil

c. Some additional items that should be carried in your pockets at all times are:

(1) Contact gloves

(2) 550 cord/Avalanche cord (10 meters)

(3) Flashlight w/ tactical lens and spare batteries

(4) Chemlights or route marking equipment

2. **ASSAULT LOAD**.

a. The Assault Load (Line 1) is equipment in addition to the basic cold weather uniform requirements, and is carried in the load bearing vest (LBV), butt pack and the pack system. This is the equipment carried for short duration missions such as security patrols or during the final assault phase. It is carried at all times when you are away from your bivouac site.

(1) An extra insulating layer (polypropylene long underwear, buffalo jacket, woolly pulley, etc.)

(2) Protective layer (ECWCS parka and trousers if not worn)

(3) LBV with 2 quarts of water and first aid kit

(4) Rations for the time away from your bivouac site

(5) Extra socks and gloves

(6) Isopor mat (strapped to assault pack or carried on the ski pole)

(7) Over-the-snow mobility (skis, poles, wax kit and/or snowshoes)

(8) Mission essential gear as required:

(a) T/O weapon w/accessories (sling, magazines, cleaning gear, bayonet/K-bar, and basic allowance of ammunition)

(b) *Extra ammunition, demolitions, and pyrotechniques as the mission dictates

(c) *Optical gear (binoculars, night vision devices, etc.)

(d) *Communications equipment (field phones, spare batteries, etc.)

(e) *Navigational equipment (map, compass, GPS, etc.)

**NOTE**: Mission essential gear items indicated with an * are spread-loaded throughout the unit as the mission dictates. Also, it may be required for designated personnel (such as RTOs) to carry the assault load in the large Vector pack vice in the small assault pack.

3. **COMBAT LOAD**. The Combat Load (Level 2) is the equipment carried for longer duration missions such as movements to contact. It is carried in the large Vector pack and consists of essential gear required in the event of an unplanned bivouac and the gear required to conduct medevacs. The following items are in addition to the items already being carried in the Assault Load:

a. Sleeping bag (w/ bivy bag if issued) inside a WP bag

b. Snow shovel (to dig expedient shelters, fighting positions, or rescue avalanche victims)

c. Individual/squad stove (Whisperlite, Peak 1, Optimus 8R, etc.)

d. Fuel bottle w/ fuel

e. Thermos

f. Poncho (for expedient shelters or medevac purposes)

**NOTE**: If the gear list dictates that each man carries the Assault Load, then 1 man per squad will also bring the Combat Load items. These items may be spread-loaded throughout the squad to prevent over-burdening 1 man with the extra weight. If all personnel are carrying the Combat Load, then items b, c, d, e, and f are 1 each per 2 men.

4. **EXISTENCE LOAD**. The Existence Load (Level 3) is any extra gear that is required that can be brought up to the forward combat elements once the situation allows. Ideally, each

fire team packs their excess gear in one sea bag, and it comes forward on the log train. It includes, but is not limited to:

a. Extra insulating layers

b. Extra socks

c. Extra glove and mitten liners

d. Toiletries

e. Sewing kit

5. **GROUP STORES** Ideally, each fire team will have their own sled to haul their group stores. The individual Marines may pull this sled during their movements, carried on the roof of their SUSV, or it may be brought up via the log train during the consolidation phase. The group stores inside the fire team sled should consist of the following:

a. 1 ECW tent complete (tent body, fly, pole set)

b. Extra fuel for the stoves

c. 1 case of extra MRE's/RCW's

d. 2 shovels and pioneer gear as required (hatchet, belay rope, etc.)

e. 2 pr. climbing skins (if ski-borne)

f. 1 whiskbroom

g. 1 team cook set

h. Candles

i. Trash bags

**NOTE**: Proper packing of the group stores is taught in the Fire Team Sled class.

6. **PACKING THE PACK**. Because most Marines are familiar with how to pack a pack, these are general guidelines only.

a. Keep your pack as compact as possible. The bulk of the weight should be next to your body, low and centered. This will assist you in maintaining your balance when skiing or snowshoeing.

b. Your shoulder and waist straps should be properly adjusted. They should be snug yet allowing freedom of movement to prevent cutting off circulation.

c. Limit the amount of gear and miscellaneous equipment you have strapped to and dangling from the outside of your pack. This will reduce the amount of snagging when moving through dense vegetation and should prevent lost gear that may save your life.

d. Ensure you properly waterproof any gear you do not want to get wet. Rain or snow can quickly soak through your pack and make your last pair of dry socks wet.

e. Use stuff sacks to keep small or related gear neatly stowed where you can readily locate it, even in the dark.

f. When not wearing your protective layer or insulating layers, keep it handy at the top of the pack. When taking a break during a movement, you will be able to quickly don a layer to prevent getting chilled.

g. Keep your stove and fuel bottle in the outside pockets of your pack. They may leak and soak your equipment with fuel if stored inside your pack.

h. Keep your ski wax kit readily available in the outside pockets, this allows easy access to your wax during short halts.

i. Keep some high-energy snacks handy in the exterior pockets of your pack. This will allow you to refuel your system on short breaks during movements without the need to dump your gear.

j. When not wearing your over whites, keep them handy under the map flap of your pack, so you can quickly change your camouflage as the terrain requires.

k. Snowshoes should be secured to your pack with straps, Para cord, bungee cord, etc. with the tails pointing up and the shovel wrapping in under the bottom of the pack.

l. Skis should be carried on the pack by sliding the tails down through the external side pockets. The tips are then strapped together with a toe strap.

**UNITED STATES MARINE CORPS**
Mountain Warfare Training Center
Bridgeport, California 93517-5001

FMST.07.0
5
10/22/01

# STUDENT HANDOUT

## MILITARY SKI EQUIPMENT

**TERMINAL LEARNING OBJECTIVE**. Given snow covered mountainous terrain, military skis, cold weather clothing and an assault load, conduct a military ski movement, in accordance with the references. (FMST.07.05)

**ENABLING LEARNING OBJECTIVES**.

(1) Given a pair of military skis with poles, maintain military ski equipment, in accordance with the references. (FMST.07.05a)

(2) Without the aid of references, select from a given list the correct definition for the parts of the military ski, in accordance with the references. (FMST.07.05b)

(3) Given a pair of military skis with bindings, adjust the ski bindings, in accordance with the references. (FMST.07.05c)

(4) Given a pair of military skis and ski waxes in prevailing snow conditions, wax skis to provide both grip and glide, in accordance with the references. (FMST.07.05d)

**OUTLINE**

1. **TYPES OF SKIS**. To understand the design of skis, we should first look at what their differences and uses are. For simplicity we can break skiing into three categories.

a. Nordic Skiing (Cross-Country). There are several different types of cross-country skis. They are as follows:

(1) Racing. This type of ski is the narrowest of the cross-country skis. The racer wants lightweight skis that do not drag in the tracks (The racer always skis in a prepared track so support is not a concern). There are two types of racing skis; the classic and the marathon skating ski.

(2) Light Touring. This type of ski is narrow to medium in size. The light tour skier wants skis suited to racing but they must be wide enough to support his weight in the powder and for off track skiing.

(3) Touring. This is a medium to wide ski. The tour skier wants support when breaking trail and skiing the powder in the backcountry where a wider ski is a must.

(4) Keep in mind that all these skis mentioned have no metal edges; therefore they are not desirable for descending steep slopes.

b. Alpine Skiing (Downhill). This is the type of skiing done at ski resorts. The alpine ski makes skiing very easy for the normal skier. It is designed to go down a slope with minimum effort, but it is nearly impossible to move on gentle rolling terrain on this gear.

c. Mountaineering Skiing. This is the widest of the ski types. The mountain skier wants skis for support and turning. A wider ski is designed to turn more easily and the increased width supports better in the backcountry. The mountaineering ski is a cross between Nordic and Alpine type skis with a metal edge and bindings that allow free movement of the heel.

2. **MILITARY SKIS**. The current military skis in the system are the Asnes double cambered skis. These skis vary in length from 180cm to 210cm and have a hole in the tips for towing. The size of the skis issued to individuals will be dependant upon their weight. Heavier Marines should receive longer skis to assist in flotation.

a.. Nomenclature of the military ski. (FMST.07.05b)
a

(1) Tip. The obvious forward point of the ski.

(2) Shovel. The curvature at the front of the ski that helps push aside the snow as it moves.

(3) Tail. This is the rear of the ski. It has a notch in the center for attaching climbing skins.

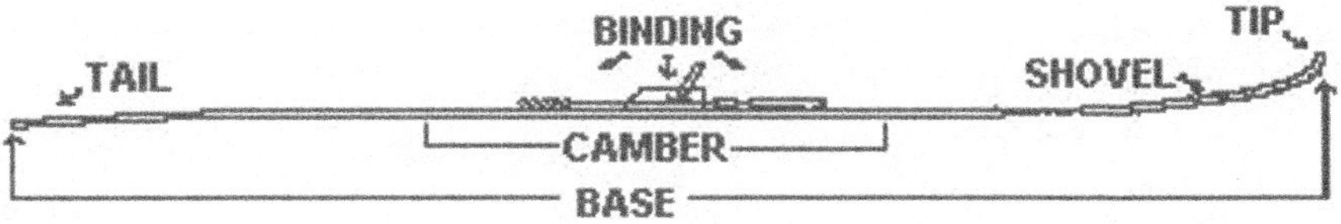


(4) Base. This is the bottom of the ski, which is made from a synthetic material called P-Tex.

(5) Camber. If you put a pair of skis together, you will notice that there is a bow in the center portion of the skis, what ski manufactures call a "camber". It is often referred to as

the wax pocket. When you put weight on the ski you will notice that the bow will flatten out. The amount of weight needed to flatten the bow depends on the skier's weight and ability.

(a) Single camber. The camber in the ski is soft so that more of the ski is in contact with the snow. This makes steering easier and provides more control.

(b) Double camber. This is a term that is used to describe the "stiffness" in the camber of the ski. A double camber ski will require more force to flatten out which causes steering to be more difficult, but the stiffer ski will have a better gliding and grip wax pocket.

(5) Binding. The NATO 120 binding is an all-metal binding consisting of a cable clamp, toe plate and a cable. This will be discussed in further detail in the next section.

b. Other features

(1) Tracking Groove. If you look at the bottom of almost every cross country ski, you will see a U-shaped or L-shaped cut called a "tracking groove" going from just below the tip down to the tail. It is designed to help the ski run over the snow in a straight line. Without the groove, the ski will tend to wobble or move from side to side.

(2) Kick Zone. This area is located within the camber and is where the wax is applied. The kick zone will vary between skiers due to his weight, ability and the terrain being skied. As a general guideline, the zone begins 6 inches before the binding and extends to 6 inches beyond the binding.

(3) Glide Zone. This is the area that remains in contact with the snow surface. The glide zone runs from the ends of the kick zone to the respective ends of the ski.

(4) Metal Edge. This is an alpine ski feature that is essential for mountaineering skis. The edge can be offset or flush with the running surfaces of the ski.

(5) Side Wall. The side covers of the ski that protects the core of the ski from warping due to water damage.

(6) Side cut (Waist). The difference in width measurements from tip to tail is referred to as the ski's side cut. The side cut makes the ski easier to turn when pressure is applied on the ski at an angle to the snow surface.

(7) Flex. There are three things involving flex:

(a) Tip flex. Soft tips follow the terrain by easily flowing over bumps, dips and irregularities in the snow. If the tip is too soft, the ski tends to wander and become difficult to control in turns. Moderate tip flex is more desirable for backcountry touring and mountain skiing, thus providing better flotation in powder and adequate control when turning.

(b) Tail flex. Tail flex is similar to tip flex in its response to snow and turning. If it is too soft, the ski may wash out or not hold an edge while turning.

(c) Torsional flexibility. This refers to the twisting action from side to side that a ski goes through while in a turn or track. A good touring or mountain ski has a torsionally stiffer tip, which gives the ski more holding power and better edge control when turning.

3. **SKI BINDING**. The NATO 120 Binding is a versatile binding because it can be fitted to a variety of boots. The Vapor Barrier (VB) boot and the 75mm box-toed leather ski march boot being the most commonly used in the Marine Corps.

a. Nomenclature of the NATO 120 Binding

(1) Toe plate. This consists of a wing nut fastener, locking lever, and two adjustable toe plates designed for proper emplacement of the toe of a boot toe.

(2) Cable clamp. This is located in the front of the binding and is designed to tension the cable around the boot. The cable clamp also has a retractable nut, which allows for two full sizes of adjustment of a cable to a boot.

(3) Cable. This is a plastic coated cable with a coil spring portion that fits behind the back of the heel. The cables come in four sizes with a different colored band representing the size rating of that cable. The color and corresponding boot size are as follows:

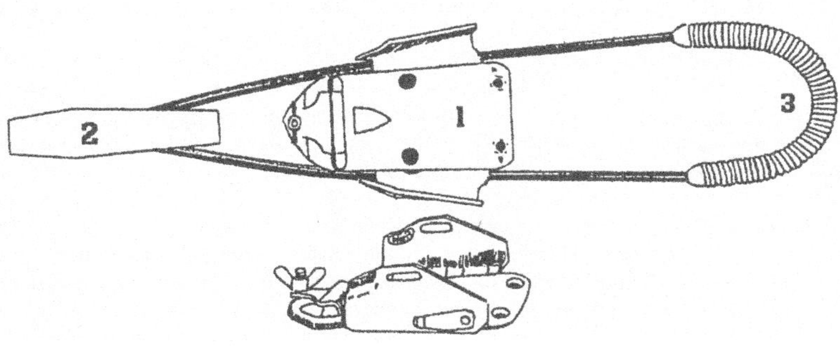

NATO 120 SKI BINDING

| CABLE SIZING CHART | | |
|---|---|---|
| **BAND COLOR** | **SIZE** | **BOOT SIZE** |
| Blue band | XLG | 12 to 14 |
| Black band | LG | 10 to 12 |
| Green and Yellow band | MED | 8 to 10 |
| Red band | SM | 6 to 8 |

**NOTE:** Blue band cables need to be ordered separately.

b. Adjustment of the NATO 120 Binding:

(1) Toe plate. Due to the different size boots used with this binding, the adjustment of the toe plate is necessary. Start by loosening the wing nut and lifting the locking lever up. Place the toe of the boot behind the wing nut ensuring that the heel of the boot is centered on the ski. Align the toe plates against the welt of the boot and push the locking lever back down into its original position. Tighten the wing nut until no movement can be felt from the toe plate. It may be necessary to mark which ski is left or right since each toe plate will be adjusted differently.

(2) Cable clamp. Ensure that the proper tension is attained and secure the clamp down. Minor adjustments to the tension can be done by unscrewing or tightening the nut. Once the cable is clamped down, there should be no lateral movement of the boot.

(3) Cable. No adjustment is needed for the cable. The cable clamp will compensate for the proper adjustment. Attention to the correct cable length for the boot size is important.

4. **MILITARY SKI POLES**. The ski poles aid the skier in movement, balance and timing.

a. Nomenclature of the Military Ski Poles.

(1) Wrist straps. The leather wrist strap should be adjusted to support the wrist for pushing while cross-country skiing. Once adjusted they should not be cut as different types of gloves or mittens will require readjustment.

(2) Handgrip. The handgrip is made of a hard plastic and is where the wrist straps are attached. Some poles are designed with detachable color-coded handgrips, which indicate that the poles can be converted into an avalanche probe pole. In order to transform the ski poles into a probe, it is necessary to have a red colored top handgrip, representing a "male" pole, and a white colored top handgrip representing a "female" pole.

(a) To convert the ski poles to an avalanche probe pole, remove the color-coded handgrips by unscrewing them from the poles.

(b) Screw the male end into the female end.

(c) Remove a basket from one end of the pole.

(3) Shaft. The shaft is made from one piece of hollow aircraft aluminum.

(4) Basket. The basket is located near the end of the pole. This basket allows the pole to remain above the surface of the snow during pole plants.

(5) Point. This is located at the end of the pole, also known as a Ferrule. This is what penetrates the snow surface during pole plants.

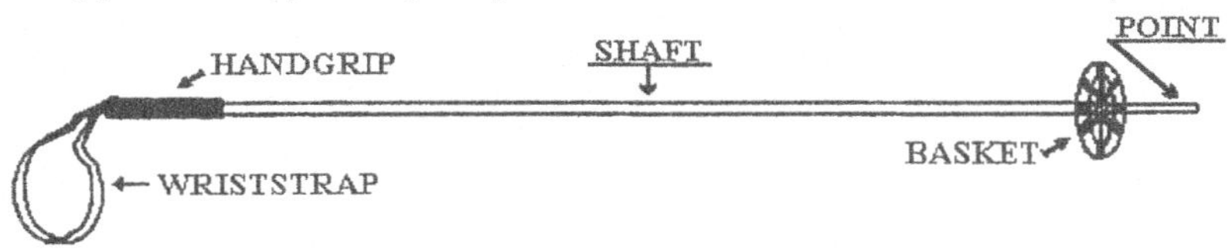


**MILITARY SKI POLE**

b. Other considerations

(1) Length. The military ski pole comes in three lengths, 130cm, 137cm and 147cm. To properly size a pole to an individual, place the tip on the deck, the handgrip should fit snugly under the individual's armpit.

(2) Weight. Each pole weighs approximately one pound.

5. **CARE AND MAINTENANCE OF THE MILITARY SKI**. Serviceability checks and proper maintenance is a regular routine for ski equipment.

a. Serviceability Checks.

(1) Breakage. Skis are vulnerable to load stress, particularly the tips and tails.

(2) Delamination. This occurs when the plastic coating separates from the ski causing water damage to the inner core. Skis should be frequently checked for nicks and gouges of the coating.

(3) Ski base. The ski base should be uniformly flat and smooth. Check for possible gouges and cuts. These may hamper the glide of the ski and create problems in waxing. Any gouges or cuts should be filled to prevent an unstable ski.

(4) Ski edges. If the ski edge is separated from the ski, turning will become a problem. Generally, if large pieces of the edges are missing, mobility will become difficult, especially when turning or edging. The ski should be replaced.

(5) Bindings. Check all metal parts for stress fractures and missing parts. Ensure that the cable is not missing large sections of the plastic coating or that the coil spring is not over stretched. This can possibly damage your boots.

(6) Detuning. If the skis are new, the metal edges are very sharp. Detuning the tips and tails approximately six inches on both sides will help prevent you from "catching an edge".

b. Care of the Ski

(1) Heat. Don't place the ski next to direct heat, because the bottom of the ski could easily melt. Anytime heat is placed near a ski, it should only be for a few seconds. Caution should be taken not to attach skis too close to the exhaust pipes on tracked vehicles.

(2) Snow/ice. Remove snow and ice from the skis before bivouacing. Icing may occur making the ski difficult to wax later on.

(3) Waxes. Remove all waxes before bivouacing, or you may be skiing with yesterday's wax on today's snow conditions.

(4) Staging skis. Never place the tails of the skis straight into the snow. This can cause damage to the tails by striking a possible hidden object i.e. rocks, tree stumps, etc. Skis should be staged during breaks by directing the base of the ski toward the sun using the poles for support. This will keep the grip wax fresh and applicable.

c. Maintenance of the Ski.

(1) Base preparation. Base glide wax is used to protect the ski's synthetic base and to maximize the forward speed of the ski. The simplest method for waxing is to rub the wax directly on the base. However; hot waxing allows the wax to penetrate deeper and last longer. Here's how it is done:

(a) Ensure that the base is clean and dry.

(b) If the base is damaged due to gouges, a P-TEX candle can be melted into the problem areas. Ensure that the excess is smoothed evenly across the base. Repeat the process if the patch shrinks below the surrounding base surface.

(c) With a hot wax iron, apply the based glide wax by dripping it over the base of the ski.

(d) Run the wax iron over the ski to reheat the wax and to spread it evenly along the base of the ski. Always keep the iron moving to avoid melting the synthetic base.

(e) Remove excess wax with a scraper to include the tracking groove.

(2) Metal edge tuning. When the ski metal edge is dulled or pitted, turning the ski can be difficult. Therefore, the edges should be sharpened as needed, except in the shovel and tail area. These two areas should be divided (detuned) for maximum turning efficiency.

(a) To file the ski edges, hold the file parallel and lengthwise against the side of the edge. Your thumb should be on top, and your fingers curled under, acting as a guide along the ski itself.

(b) File from the tip to tail, while being careful to keep the file at a 90-degree angle to the base; avoid excessive filing. To smooth out the operation of the ski, the first 6 inches at the tip, and the last 3 inches at the tail, should be slightly dulled.

6. **SKI WAXING**

a. Factors Affecting Wax. Wax is applied to the base of the skis to prevent slipping, influence momentum, and help maintain glide. The goal of waxing is to find the proper wax combination for optimum grip and glide, without sacrificing either. As the snow conditions and temperature change, the wax required will also change.

b. Ski Waxes

(1) Each wax has its range of ideal snow condition. The type of snow (wet or dry) and the temperature will play a role in what wax you choose and how you will apply it. The wax chart below provides examples of some waxes commonly used.

**WAX CHART**

| **WAX** | **SNOW TEMPERATURE (°F)** | **USAGE** |
|---|---|---|
| Polar | -22° to 5° | Very cold snow condition wax. Frequently used in the polar regions of Europe and North America. |
| Green Special | 5° to 12° | Cold snow condition wax and ideal for those who want a long kick zone. |
| Green | 9° to 19° | Cold snow wax. |
| Blue Special | 16° to 23° | Cold snow wax with a wide range. |
| Blue | 18° to 27° | Used on moderate cold snow. |

| | | |
|---|---|---|
| Blue Extra | 19° to 32° | Demands new snow and low air humidity (below 45-50%). |
| Violet Special | 30° to 32° | When Blue Extra becomes somewhat slippery, a thin layer of Violet Special probably is the right wax. |
| Violet | 32° | To be used when the snow is in a stage going from cold to wet, around freezing (32°). |
| Violet Extra | 32° to 34° | Used when the snow is moist. |
| Red Special | 30° to 36° | Wax for moist, new snow, but might go into some colder snow and ball up the snow in your kick zone. |
| Red | 32° to 38° | Wax for moist to wet new snow, when making snowballs is very easy. |
| Red Extra | 34° to 38° | For wet, new snow and glazed tracks. Must be applied in an even layer to prevent balling. |

**KLISTER CHART**

| **KLISTER** | **SNOW TEMPERATURE (°F)** | **USAGE** |
|---|---|---|
| Green | -13° to 27° | To be used as a first layer as a binder for other klisters or hard waxes in very abrasive conditions. Used alone for icy conditions at very cold temperatures. As a first layer, it should be heated into the base of ski. |
| Blue | 5° to 32° | For frozen, icy conditions. Can also be used as a base klister for wet-snow klisters in improve wear. |
| Violet | 25° to 37° | Use for snow on both sides of freezing, but also when conditions are changeable and mixed with fine-grained snow. |
| Red | 32° to 41° | Used on changing, coarse-grained snow on both sides of freezing. Best on the warm side of freezing when the snow is wet. |
| Orange | 36° to 52° | Used when the snow has a high water content such as slush and the air temperature is well above freezing. |

(2) Two wax system. Instead of carrying around all the above-mentioned waxes, we can keep it simple using a two-wax system. This is what training units will be issued at MWTC.

**TWO WAX SYSTEM**

| WAX | SNOW TEMPERATURE (°F) | USAGE |
|---|---|---|
| Blue | 5° to 32° | A wide range wax for normal cold snow. |
| Red | 32° and warmer | Ideal for conditions around freezing and slightly warmer. |

c. Wax and its effects on snow. When wax is spread on the ski's base, it provides a cushion for thousands of tiny snowflakes to stick into. For a brief moment, when all the skier's weight is on one ski, a multitude of snow crystals are embedded in the wax, holding the ski firm while the skier pushes (or kicks) off. As the ski begins to slide forward, the pressure release and friction of the sliding ski releases this bond. A thin, microscopic layer of water (as a result of friction between the ski's base and the snow) causes the ski to glide until it stops and downward pressure is again applied.

d. Classifications of Snow

(1) Fresh fine-grain snow. This usually occurs after a snowstorm when temperatures are cold. Each flake has a very distinctive feature, like the fingers of an outstretched hand.

(2) Coarse-grain snow. This develops one to three days after a snowstorm. This snow undergoes a transformation in which the crystals begin to lose their shape. Evaporation and compression cause a rounding of the fresh snow's sharp points, like taking an outstretched hand and starting to form a fist.

(3) Granular snow. This type of snow develops when the temperature rises above freezing and the snow melts and refreezes. It is similar in shape to a closed fist, with no relationship to the original snowflake.

(4) Wet snow. This type of snow is usually found during the spring. It may also occur at other times, particularly in regions of moderate climate. This type of snow can be made into a heavy, solid snowball. In extreme conditions, wet snow will become slushy, and contain a large amount of water.

(5) Dry snow. This type of snow is generally associated with winter at its height, but it can occur in late autumn, as well as in the spring, when abnormally low temperatures occur. This snow is light and fluffy. It cannot be compressed into a hardball unless the snow is made moist by holding it in the hand. At extremely low temperatures, such as those found in the far northern regions, this snow is like sand, and has very poor sliding qualities.

e. Proper Selection and Application of Ski Waxes. Grip waxes are formulated to provide optimum gliding and gripping characteristics for various types of snow and temperatures. Each type is labeled with appropriate instructions on its intended use, i.e., wet; moist or dry snow conditions, or temperature. Since the type of wax varies between manufacturers, no particular type of wax can be prescribed for each classification of snow; however, the instruction on each container specifies the weather conditions and the type of snow where performance of the wax is best. To provide a proper grip, varying amounts, combinations, and methods of application of different waxes may be used.

(1) Tips for apply grip wax:

(a) Wax in the temperature being skied.

(b) Ski should be dry when waxing.

(c) The wax should be corked in from tip to tail.

(d) Grip waxes may be crayoned, and then corked in.

(e) When in doubt start with a harder wax.

(f) Don't apply a harder on top of a softer wax.

(g) Several thin layers work better than one thick layer.

(h) During movement carry wax in an inside pocket to keep it warm.

(i) Do not put newly waxed skis on the snow, until the wax has cooled to air temperature. If the lead group has a specific wax on for the conditions of starting out, the trailing group might want to use a warmer or colder wax due to different track conditions left by the lead group.

f. Test. Before you move out on a ski march or any other type of movement, test the waxing job prior to movement and re-wax if necessary. It will normally take several hundred feet of skiing for the wax to function properly.

g. Wax Kit. The wax kit should include:

(1) Two grip waxes (The two wax system)

(2) Cork

(3) Scrapper

7. **CLIMBING SKINS**. Climbing skins were originally made from sealskin. Modern day skins are made from synthetic fur called "mohair". These mohair strips attach to the bottom of the skis, which allow the skier to slide forward, but not back. Used for ascending moderate to steep terrain or pulling heavy loads over long distances.

(1) Nomenclature of the climbing skin.

(a) Skin. This is two sided. The adhesive side is placed against the base of the ski. The mohair side is the portion, which comes in contact with the snow.

(b) Heel Clamp. This secures the skin to the tail of the ski.

(c) Toe clamp. This secures the skin to the tip of the ski. It is normally equipped with a rubber-tensioning device.

CLIMBING SKIN

(2) Fitting. The M-buckle located on the toe clamp is held in place by inserting the actual skin through the buckle and folding in back of itself. To adjust skins, simply pull apart the folded adhesive apart and move the clamp forward or back. When the correct length is attained, fold skin back on itself.

(3) Maintenance and storage. To maintain the adhesive side after each use, the skins must be air-dried. To store when not in use, find the midpoint and fold the two adhesive sides back on themselves, then store in carrying bag.

**UNITED STATES MARINE CORPS**
Mountain Warfare Training Center
Bridgeport, California 93517-5001

FMST.07.0
6
10/22//01

**STUDENT HANDOUT**

**MILITARY SNOWSHOE MOVEMENT**

**TERMINAL LEARNING OBJECTIVE**. Given snow covered mountainous terrain, military snowshoes and an assault load, conduct a military snowshoe movement, in accordance with the references. (FMST.07.06)

**ENABLING LEARNING OBJECTIVES**.

(1) Without the aid of references, select from a given list the advantages of snowshoes, in accordance with the references. (FMST.07.06a)

(2) Without the aid of references, select from a given list the disadvantages of snowshoes, in accordance with the references. (FMST.07.06b)

(3) Without the aid of references, select from a given list the correct description of the parts of a military snowshoe, in accordance with the references. (FMST.07.06c)

(4) Given a pair of military snowshoes, adjust the military snowshoe bindings, in accordance with the references. (FMST.07.06d)

(5) Given a pair of military snowshoes in a snow covered mountainous environment, demonstrate snowshoe techniques, in accordance with the references. (FMST.07.06e)

**OUTLINE**.

1. **ADVANTAGES AND DISADVANTAGES OF SNOW SHOEING**

a. Advantages of Snowshoes. (FMST.07.06a)

(1) Training. Little training time is required to gain a high degree of proficiency in their use.

**6**

(2) Maintenance. Little maintenance is required.

(3) Heavy loads. Carrying and pulling of heavy loads on gentle terrain is relatively easy.

(4) Confined areas. Movement in confined areas and around equipment is relatively easy. Snowshoes are particularly useful for individuals working in confined areas, such as bivouac sites and supply dumps. Snowshoes are also helpful to drivers, gun crews, cooks and other support personnel.

b. Disadvantages of Snowshoes. (FMST.07.06b)

(1) Rate of movement. The rate of movement for a unit is extremely slow, and inefficient, in terms of energy expended.

(2) Moderate to steep slopes. Movement on moderate to steep slopes is extremely difficult.

(3) Thick or cut-off brush. Movement through thick or cut-off brush is difficult.

(4) Fire and movement. Quick movement, as needed during fire and movement is difficult.

2. **SNOWSHOE MOVEMENT RATES** (Reduce by one-third for mountainous terrain).

| MOVEMENT MODE | UNBROKEN TRAIL | BROKEN TRAIL |
|---|---|---|
| On foot with less than 1 foot of snow | 1.5-3 KPH | 2-3 KPH |
| On foot with more than 1 foot of snow | .5-1 KPH | 2-3 KPH |
| Snow shoeing | 1.5-3 KPH | 3-4 KPH |

3. **TYPES OF SNOWSHOES**. There are three basic types of snowshoes.

a. Magnesium. The magnesium snowshoe is the lightest and most durable of the three types. The nylon binding used with this snowshoe is adaptable to all types of issued footwear. The magnesium snowshoe also has teeth under the sides, which are intended to aid traction.

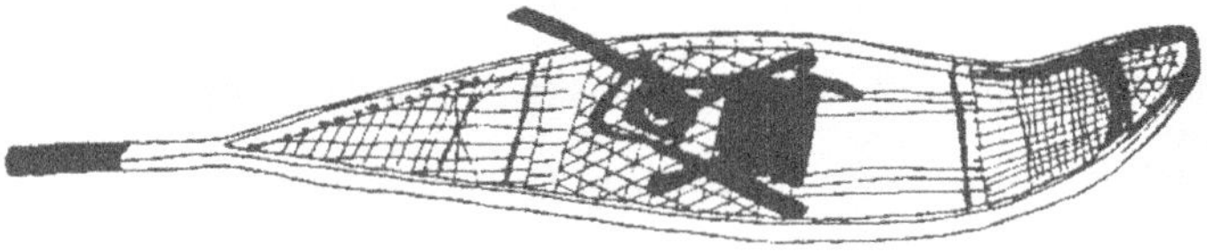

MAGNESIUM SNOWSHOE

b. Assault. This type of snowshoe is short, wide and oval in shape, with no tail. It is best utilized when working near equipment and heavy weapons, due to it offering little floatation.

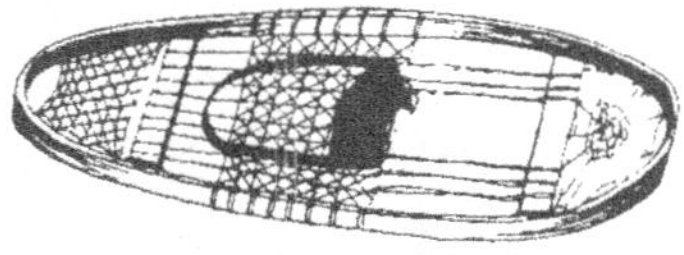

ASSAULT SNOWSHOE

c. Improvised Snowshoes. Improvised snowshoes may be constructed by forming a frame from green, flexible branches, then weaving string, 550 cord, wire or branches to form a supporting surface. Tying branches can also make a satisfactory pair from thick fir, or spruce trees, to one's feet. A very simple Sasquatch binding can be made with a short length of cord or wire.

IMPROVISED SNOWSHOE

4. **NOMENCLATURE OF THE MILITARY SNOWSHOE**. The magnesium snowshoe consists of nine parts: (FMST.07.06c)

a. Tip. This is the front portion of the snowshoe frame.

b. Tail. This is the back portion of the snowshoe frame.

c. Binding. This is constructed of a nylon material and fits the boot.

d. Crossbars. There are two crossbars welded to the frame to reinforce it.

e. Window. This is the opening in the snowshoe, which allows the toe of your boot to pivot through.

f. Webbing. Made from galvanized aircraft cable covered with nylon.

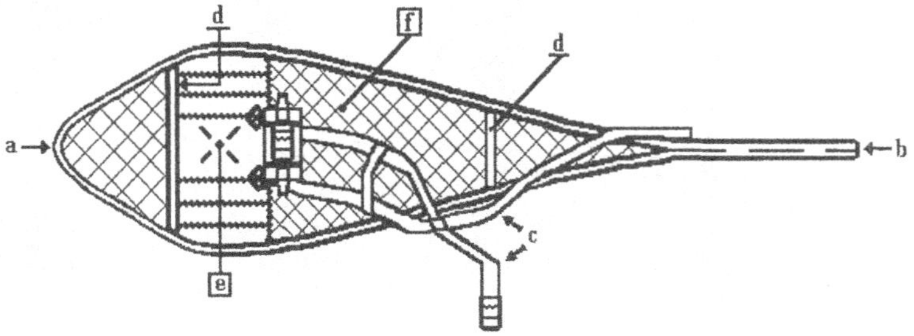


g. Shovel. Front of the snowshoe, which has a shovel like resemblance.

h. Frame. Consists of a magnesium alloy.

i. Teeth. These are located on the underside of the frame and create more traction on the surface when worn.

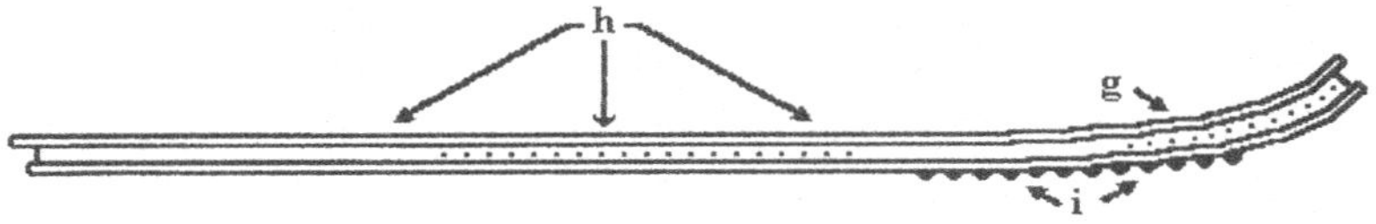


5. **CARE AND STORAGE**

a. Webbing and Bindings. Check load bearing webbing and bindings frequently.

b. Magnesium Frames. Check frames of magnesium snowshoes for stress fractures.

6. **BINDING ADJUSTMENT**. Proper snowshoe binding adjustment will ensure that:

a. Ball of Foot. The foot pivots freely about the ball of the foot, so that the toe of the foot moves through the window of the snowshoe.

b. Heel of Foot. The heel of the foot is centered on the snowshoe.

PROPER SNOWSHOE BINDING ADJUSTMENT

c. Fit. The binding fits snugly to provide adequate control, but not so tightly that circulation in the feet is impaired. A sloppy fitting snowshoe will make movement extremely difficult.

d. Toe strap. May be used to secure the binding together, to help limit the number of blown bindings.

## 7. TECHNIQUE

a. General. There is little difference in snow shoeing compared to normal walking, except that the surface being walked on is inconsistent, and snowshoes are longer, wider, heavier, and consequently more awkward, than normal footwear. With standard military snowshoes, the stride is somewhat longer than in normal walking, but the shape of the snowshoe allows the snowshoer's stance to be a normal width, thereby reducing much strain and fatigue on his hips and legs. It should be stressed that the snowshoer should walk in a relaxed, and normal rolling toe manner, and should only lift the snowshoe high enough to clear the surface of the snow.

b. Turning

(1) The kick turn. This is normally the easiest way to change directions on level ground. One snowshoe is swung up to the front so that its tail is on the snow, then it is allowed to pivot toward the new direction. The other snowshoe is then brought around.

KICK TURN

(a) On steep terrain. It is important to remember to step off with the uphill foot, when changing direction. For example: If making a turn to the right, shift your weight to the left foot, face down the slope, and swing the right snowshoe around to point in the direction of the next switchback. Then stamp the right snowshoe into the snow. Make sure the tail is not on the left snowshoe. Now, gently shift your weight to your right foot and swing the left snowshoe around so it is parallel with your right snowshoe.

(b) Each succeeding man. When using the kick turn technique on steep terrain, try and stay well above your previous trail. This trail has undermined the snow on which you are now building the turn. As each succeeding man uses the turn, it will tend to slough off on the shoulders, and the men toward the end of the column will have a hard time getting around. This can be prevented if each man in placing his snowshoes precisely where those in front of him have placed theirs uses care. If there is only one way around an obstacle, this can be very important.

(2) The star turn. This can also be used to change direction by simply executing a series of half facing movements.

(3) Choosing a route. When climbing, plan to use the gentlest places on a slope for turns. Look ahead, and pick the route and use the terrain to your advantage. Avoid the steep parts, and don't hesitate to make short switchbacks.

c. Side Step. This is used when the slope is at a critical angle.

d. Herringbone. This is used when the slope is at a gradual angle.

e. Crossing Obstacles. Here are a few simple rules to remember:

(1) Always step over obstacles. Do this to avoid damaging snowshoes and losing balance.

(2) Never bridge a gap. Never do this with your snowshoe so that the tip and tail are higher than the center.

(3) Shallow snow. In shallow snow, there is a danger of catching, and tearing the webbing on tree stumps, or snags, which are only slightly covered.

(4) Wet snow. This will frequently ball up under the feet, interfering with comfortable walking. This snow should be knocked off as soon as possible.

(5) Deep Snow. Breaking trail in deep snow uses a lot of energy. Frequent change of the lead man should be stressed.

(6) Water. Stepping into water with snowshoes can form ice, to which significant amounts of snow can cling, making the snowshoe very heavy.

**UNITED STATES MARINE CORPS**
Mountain Warfare Training Center
Bridgeport, California 93517-5001

FMST.07.08
10/22/01

**STUDENT HANDOUT**

**PREVENTIVE MEDICINE AND WATER PURIFICATION**

**TERMINAL LEARNING OBJECTIVE.** Given a unit in a cold weather environment and the necessary equipment and supplies, perform cold weather preventive medicine, in accordance with the references. (FMST.07.08)

**ENABLING LEARNING OBJECTIVES.**

1) Without the aid of references, from a given list select the five important areas of personal hygiene, in accordance with the references. (FMST.07.08a)

2) Without the aid of references, from a given list select the correct steps taken to ensure proper immunization prior to deployment, in accordance with the references. (FMST.07.08b)

3) Without the aid of references, from a given list select the three methods of water purification, in accordance with the references. (FMST.07.08c)

4) Without the aid of references, from a given list select the four forms of halogens used for water purification, in accordance with the references. (FMST.07.08d)

5) Without the aid of references, from a given list select the definition of giardia, in accordance with the references. (FMST.07.08e)

6) Without the aid of references, from a given list select the most common drug used to treat giardia infection in the United States, in accordance with the references. (FMST.07.08f)

7) Without the aid of references, from a given list select the three important reasons for proper waste disposal, in accordance with the references. (FMST.07.08g)

**7**

8) Without the aid of references, from a given list select the correct methods of waste disposal, in accordance with the references. (FMST.07.08h)

1. **PERSONAL HYGIENE**. Your body's defense mechanisms break down due to the physical demands of the arduous terrain, environment, stress and metabolic changes. Personal hygiene becomes the key to prevent the spread of communicable diseases. Five important areas are: (FMST.07.08a)

a. Body.

(1) The body should be washed frequently in order to minimize the chances of small cuts and scratches developing into full-blown infections and as a defense against parasitic infections.

(2) A daily bath or shower consisting of soap and hot water is ideal. However, when this is not possible, you should:

(a) Give yourself a "sponge bath" using soap and water, making sure particular attention is given to body creases, i.e., armpits, groin area, face ears, hands and feet.

(b) If water is in extremely short supply, you should take an "air bath" by:

(1) Remove all clothing and hang it up to air.

(2) Expose the body for two hours to the sunlight, which is ideal, but the effects will basically be the same if done indoors or during an overcast day. BE CAREFUL NOT TO SUNBURN!

(3) Shaving. Shaving at night before bedtime is preferable because shaving will remove facial oils, which aid in protecting the face from a harsh cold environment during the day.

b. Hair.

(1) Clean frequently.

(2) Inspect at least once a week for parasites.

c. Hands. Hand washing with soap and water will cut down in the transmission of viral and parasitic infections via the person-person and fecal-oral route.

(1) Keep fingernails trimmed and clean to prevent accidentally scratching yourself and to prevent a harborage for bacteria.

d. Feet. Your prime source of transportation, so take care of them!

(1) Inspect frequently for:

(a) Blisters.

(b) Infections, bacterial and fungal.

(2) Preventive measures:

(a) Keep your feet dry, change socks frequently.

(b) Use antifungal foot powder.

(c) Antiperspirants. Pretreatment for 2 weeks prior to field with an antiperspirant containing Aluminum Chlorhydrate. This will aid in the prevention of athlete's foot.

e. Oral Hygiene. The mouth and teeth should be cleaned at least daily to prevent tooth decay and gum disease.

(1) Ideally with:

(a) Tooth brush.

(b) Tooth paste.

(c) Dental floss.

(2) If these items are not available:

(a) A "chew stick" made from a clean twig about 8" long and finger width. Chew one end until it becomes frayed and brash-like, then use as a toothbrush.

(b) Field expedient dental floss can be pulled from the centerlines of paracord.

(c) Rubbing them vigorously with a clean index finger should stimulate the gums at least once daily.

2. **IMMUNIZATIONS (NAVMEDCOMINST 6230.3)**. In addition to being required, immunizations are your best front line defense against communicable disease.

a. Eight weeks prior to deployment: (FMST.07.08b)

(1) Ensure Basic Series Completed (BSC) on required immunizations.

(2) Determine if additional immunizations/chemoprophylaxis will be required in that geographical area of operations.

(3) Identify and flag health risk personnel: i.e. TB converters (NAVMEDCOMINST 6224.1), hepatitis (NAVMEDCOMINST 6230.1A), mononucleosis, parasitic infections and anemic personnel.

(4) Review medical intelligence reports:

(a) NEPMU, Navy Environmental Preventive Medicine Unit.

(b) G-2, USMC.

(c) NAVDISVECTOECOLCONCEN, Navy Disease Vector Ecology and Control Centers.

(d) CDC - Center For Disease Control.

(e) WHO - World Health Organization.

(f) Internet travel health sites.

3. **WATER.** A satisfactory source is one where quantity and quality is enough to supply the needs of all the troops. Water or ice taken from the environment should be considered contaminated and must be purified. Be aware that cold weather slows the chemical reaction time of purification measures.

a. Sources:

(1) Rivers.

(2) Streams.

(3) Ponds.

(4) Lakes.

(5) Ice.

(6) Snow.

b. Water Source Selection. The choice of a water source is influenced by:

(1) Whether it is not contaminated by: (Check all to one mile upstream).

(a) Sewage.

(b) Biological.

(c) Enemy pollution.

(d) Radiological.

(e) Chemicals.

**NOTE:** Check condition of vegetation or for dead animals.

(2) Quantity.

(3) Ease of procurement.

(4) Ease of purification.

(5) Freedom from turbidity.

(6) Excessive organic or non-organic contamination.

(7) Should be easily protectable.

4. **PURIFICATION.** Simply consists of removing or destroying enough impurities to make water safe and pleasant to drink.

a. The first step in purifying is to select a treatment method. (All figures are based on one quart of water.) The three different purifying methods are: (FMST.07.08c)

(1) Boiling. The standard recommendation is to bring the water to a rolling boil for 1 minute at sea level for complete sterilization of the water. Recent research showed that additional boiling time for sterilization was not necessary and that boiling for two minutes at any altitude rendered water safe for consumption.

(2) Halogens. The number of tablets and contact time must be considered. All charts based on one-quart canteen. (FMST.07.08d)

(a) Iodine tablets.

| TYPE WATER | TABLETS | CONTACT TIME |
|---|---|---|
| CLEAR | 1 | 25 MIN |
| CLOUDY | 2 | 25 MIN |
| COLD | 2 | 25 MIN |

(b) Chlorine bleach. The strength of the solution, number of drops and contact time must be considered.

| PERCENT | CLEAR WATER | CLOUDY WATER | COLD WATER |
|---|---|---|---|
| 1% | 10GTTS/30 MIN | 20GTS/30 MIN | 10GTTS/60MIN |
| 4-6% | 2GTTS/30 MIN | 4GTTS/30 MIN | 2GTTS/60 MIN |
| 7-10% | 1 GTTS/30 MIN | 2GTTS/30 MIN | 1 GTTS/60 MIN |

(c) Iodine solution. The strength, number of drops and contact time must be considered.

| PERCENT | CLEAR WATER | CLOUDY WATER | COLD WATER |
|---|---|---|---|
| 2% | 5GTTS/30 MIN | 10GTTS/30 MIN | 8GTTS/60 MIN |

(d) Betadine solution.

| CLEAR WATER | CLOUDY WATER | COLD WATER |
|---|---|---|
| 8GTTS/30 MIN | 16GTTS/30 MIN | 8GTTS/60 MIN |

(1) Filtration. Most filters are not small enough to completely disinfect water. They remove particulate debris, which allows lower halogen dose and improved appearance and taste of "dirty" water. Filters clog quickly if the water is dirty or has a lot of suspended particles. NEVER USE FILTERS AS YOUR SOLE METHOD OF WATER PURIFICATION.

(a) Maximum Filter Pore Size:

1) Parasitic eggs and larvae: 20 micrometers (um).

2) Giardia, E. histolytica: 5 um.

3) Enteric bacteria: 0.2 um.

4) Viruses: 0.01 um (too small for field water filtration).

5. **GIARDIA: FROM EPIDEMIC TO ENDEMIC.**

a. Definition. (FMST.07.08e) An intestinal flagellate protozoa that can cause clinical illness in humans.

(1) Found worldwide in tropical and temperate areas, the infectious cysts are passed by humans and animals and can be transmitted by direct fecal-oral route or waterborne contamination (cysts maintain viability for months in cold water). Due to these factors, Giardia is both a common cause of diarrhea in travelers to foreign countries, and campers in the United States where it has become endemic and is now considered the most common human parasite. (4% of stools examined at state public health laboratories were infected with the cysts or organism).

(2) It is found with increasing frequency in formerly "pristine" mountain streams, rivers and lakes and outbreaks have occurred in remote recreational areas and resorts throughout the western mountain states.

(3) Giardiasis is currently a risk for anyone drinking untreated surface water in the western mountains and for communities with water treatment techniques inadequate for eliminating the cysts.

b. Pathophysiology. The organism has two forms: Trophozoite and cyst.

(1) These may be passed in the stool of an infected person. The trophozoites die within one hour outside the body, however the cysts are very hardy, retaining infectivity more than two months in cold water. If the cysts are ingested, excystation is encouraged by stomach acid, and the motile trophozoite attaches by suction disc to the wall of the duodenum and proximal jejunum, where they may cause symptoms of illness.

(2) Water, ice, or food prepared with contaminated water, are the major sources of infection, but direct person-to-person transmission can occur from sexual contact or hand-to-mouth contamination.

(3) Incubation time: 5 - 25 days (7 - 10 medium). Most infections are thought to be asymptomatic carrier states, however, a wide array of clinical syndromes may occur.

c. Symptoms.

(1) A number of people have abrupt onset of explosive diarrhea accompanied by abdominal cramps, foul gas, vomiting, fever and malaise. This commonly lasts three to four days before developing the more common subacute syndrome. In most individuals, the onset is more insidious, and symptoms are persistent or recurrent.

(2) Stools become mushy, greasy and malodorous. Over a period of weeks, watery diarrhea may alternate with soft stools and even constipation. Upper GI symptoms accompany stool changes, mid and upper abdominal cramping, substernal burning, acid indigestion, sulfurous belching, nausea and foul flatus, distention with an early "full" feeling are all typically increased after a meal.

(3) Symptoms of anorexia, fatigue and weight loss are common, but fever and vomiting are infrequent except during an acute onset of illness.

d. Laboratory Parameters.

(1) The "typical" victim should suggest a diagnosis of giardiasis. This individual may have a history of camping, vacationing at a mountain resort, or foreign travel. Symptoms of upper abdominal discomfort, foul gas and soft stools arose one or two weeks after a short trip or during a lengthy one and have fluctuated for weeks. Sometimes improving, the individual held off visiting the doctor, only to have the symptoms recur. Work or school is not interrupted, however, physical discomfort and embarrassment from the gas are concerns.

(2) Laboratory confirmation of giardiasis can be difficult. Stool samples are the first means of diagnosis. Trophozoites may be found in fresh, watery stools, but disintegrate rapidly. Cysts remain in fresh stools for at least 24 hours. Cysts passage, however, is extremely variable and not related to clinical symptoms. Three samples, taken every other day, should be examined for a definite diagnosis.

(3) Average duration of symptoms has varied from three to ten weeks. Acquired resistance has been shown. Immunity is suspected, but not proven and human reinfection definitely occurs.

e. Treatment. (FMST.07.08f) Complete cure of most Giardia infections can be achieved with one of several drugs, however, no drug is effective in all cases. Treatment failure requires use of a different medication and in resistant cases, longer courses of two drugs taken concurrently has been suggested. Tinidazole (Fasgyn) has been reported 90-100% successful in a single dose of two grams orally, making it the new drug of choice; however, it has not been released for use in the United States. *Quinacrine (Atabrine, Mepacrine)* has been considered the first drug of choice in adults (dosage: 100mg tid for 5-7 days), but *Metronidazole* (flagyl) is most commonly used in the United States dosage: 250mg TID for 5-7 days.

f. Prevention.There is no safe or effective chemoprophylactic drug and travelers and campers should practice prevention. The purity of American wild waters can no longer be assumed and in developing countries, even tap water should be suspected. In addition to choosing a technique for individual field water disinfections, individuals must become more responsible for prevention of water contamination. This requires attention to personal latrines and camper trailer waste disposal, a subject that most people do not want to touch.

6. **WASTE**. The importance of proper waste disposal cannot be over emphasized.

a. It serves to: (FMST.07.08g)

(1) Eliminate harborage for rodents and vermin.

(2) Prevents attraction of rodents and vermin.

(3) Prevents a pathogenic contamination source.

b. Two basic types of waste are organic and non-organic. The basic ways to dispose of them are burning, burying, or hauling it away. (FMST.07.08h)

(1) Organic wastes:

(a) Human waste - Burn or haul away to a designated waste pit area.

(b) Urine - Use only assigned marked areas.

(c) Edible garbage - Burn. It should not be left exposed for animals, vermin or the enemy.

(d) Paper -Burn.

(e) Contaminated bandages - Sterilize/burn.

(2) Non-organic waste:

(a) Metals - Haul away or bury.

(b) Liquids - Burn or bury.

**UNITED STATES MARINE CORPS**
Mountain Warfare Training Center
Bridgeport, California 93517-5001

FMST.07.09
10/22/01

**STUDENT HANDOUT**

**WILDERNESS PATIENT ASSESSMENT**

**TERMINAL LEARNING OBJECTIVE.** Given a simulated casualty in a cold weather environment, conduct a patient assessment, in accordance with the references (FMST.07.09)

**ENABLING LEARNING OBJECTIVES.**

1) Without the aid of references, from a given list, choose when CPR may be legally discontinued, in accordance with the references. (FMST.07.09a)

2) Without the aid of references, from a given list, choose which situations CPR should not be started in a remote wilderness environment, in accordance with the references. (FMST.07.09b)

3) Without the aid of references, in the proper sequence perform a patient assessment, in accordance with the references. (FMST.07.09c)

**OUTLINE.**

1. **INTRODUCTION.**

    a. An organized prioritized assessment of a wilderness trauma victim is essential to proper care. A modified Advanced Trauma Life Support (ATLS) assessment system will be introduced in this chapter.

2. **TRIAGE.**

    a. Two situations of triage may exist in a wilderness environment.

(1) The number of patients and the severity of their injuries do not exceed the ability of the care-giver and his/her equipment to render care. In this situation, treat life-threatening problems first.

(2) The number of patients and the severity of their injuries exceed the capacity of the care-giver and his/her equipment to render care. In this situation, those with the greatest chance of survival are treated first.

3. **CPR: URBAN vs. WILDERNESS SPECIFICS**.

a. In a remote wilderness situation where advanced life support assistance may be hours or days away, CPR probably has no chance of success. In addition, administering CPR under wilderness conditions may put group members in serious danger, due to physical hazards and the effects of exhaustion.

b. Based on the above, several suggestions regarding CPR in remote wilderness areas can be made. Even though the following suggestions seem reasonable, they should be viewed as recommendations, without legal force, and not doctrine. At the present time, a rescuer who begins CPR is legally obligated to continue unless one of the following conditions is fulfilled:

(1) CPR may be discontinued legally when: (FMST.07.09a)

(a) The patient revives.

(b) ACLS is started by higher authority.

(c) Relieved by another rescuer.

(d) Too exhausted to continue.

(e) Pronounced dead by proper authority.

(2) In a remote wilderness environment, CPR should not be started under any of the following situations: (FMST.07.09b)

(a) The patient is in cardiac arrest caused by trauma.

(b) The patient is a drowning victim who has been immersed for over an hour.

(c) The patient is in cardiac arrest and advanced life support is more than an hour away, especially if the patient must be carried out.

(d) The patient's cardiac arrest was unwitnessed and the time of onset is unknown.

(e) The patient appears to be dead, based on rigor mortis, lethal injuries, or a body temperature below 60°F.

(f) Giving CPR would be hazardous to rescuers.

(3) After 30 minutes of CPR with no signs of life, further CPR is probably useless and may reasonably be discontinued. Administering CPR to a victim who is being evacuated by litter or sled is very difficult, if not impossible. Unless an ambulance or helicopter can be brought in rapidly, chances of survival are slim.

(4) Exceptions include patients in cardiac arrest caused by hypothermia; patients with another illness or injury complicated by hypothermia (avalanche burial and near drowning in cold water), and patients in cardiac or respiratory arrest caused by lightning injury. In these cases, the outlook is probably more favorable and CPR should be aggressively administered.

4. **PATIENT ASSESSMENT.** (FMST.07.09c) There are different considerations to think about in a wilderness setting. The environment plays a major role in treatment, care and even the extraction of the patient. Not only do we have to be mindful of the enemy situation, by keeping tactical. We also have to take into consideration the elements. Snow, rain, ice, and the terrain can and will pose a great danger to the patient and the rescuers. We must protect the patient from the elements. We can do this by placing an insulating barrier between the patient and the ground. Also exposing one body part at a time and recovering as you move on will help to fight hypothermia.

A) SCENE SIZE-UP.

1) Determine scene safety:

a. Rescuers: Your safety is priority. You are no good to the team as a casualty.

b. Patient: Protect the patient from further injuries, enemy, and the elements.

c. Bystanders: They can cause many accidents. Limit the personnel on scene.

d. Tactical: Have the Marines maintain security and find cover for the rescuer and the patient.

e. Fire: Forest fires can pose a formidable threat.

f. Electrical: Make sure to shut off electricity first. Careless and hasty actions injure many rescuers.

g. Special Rescues: Swift water and high angle rescues are very dangerous and trained professionals should only attempt the rescue.

2) Determine the mechanism of injury / nature of illness:

a. Obtain SAMPLE and Mechanism of Injury (MOI) history from patient, witnesses, or other caregivers.

S = Signs and/or Symptoms.
A = Allergies.
M = Medications.
P = Past Illnesses.
L = Last Meal.
E = Events of Injury.

MOI = what, when, where, and how injury occurred.

b. Environment: Survey the area to get clues. (I.e. falling rocks, next to a cliff, broken branches, etc...)

c. Bystanders: Ask them what they saw.

3) Determine number of patients:

a. Survey area for number of patients.

b. Look for the hidden patients. Patients could be buried in rock, sand or snow. They also could be spread out in bushes and trees. Be sure to have people search and identify all patients, this includes looking for patients in a crowd of bystanders.

4) Request additional help / Organize help if needed.

a. Medical assistance: Due to the lack of medical personnel and equipment, you might need to request additional help.

b. Technical assistance: When the scope of care is over your ability or higher than the CASEVAC corpsman can provide, you should request higher trained personnel.

c. Organize help at the scene: Use the people around you for help.

5) Consider stabilization of the spine:

a. Falls from 10 feet or higher, high speed collisions, unwitnessed loss of consciousness, etc...

b. Designate an assistant: Use the bystanders / marines. Tell them exactly what you want them to do and show them how to do it.

c. Emphasize and improvise control: If you are alone you can use the patient's boots, sand bags, rocks, etc as forms of temporary control.

B) <u>INITIAL ASSESSMENT</u>.

1) Make general impressions of the patient: What can you obtain from the patient's condition as you walk up? Is the patient talking, or unconscious?

2) Determine responsiveness / level of consciousness.

a) Use AVPU.

b) Assess pupils for size, equality, and reactivity.

3) Determine chief complaint / Apparent life threats.

4) Assess Airway and breathing.

a) Assessment of airway.

- Perform chin lift or jaw thrust maneuver, with C-Spine precautions.
- Clear the airway of foreign bodies.
- Insert an oropharyngeal or nasopharyngeal airway.
- Use of other airways as needed, and as equipment and training allow.

b) Assessment of breathing.

- Expose the neck and chest; assure immobilization of the head and neck.
- Determine the rate and depth of respirations.
- Auscultate the chest bilaterally. (Ear to chest wall, if necessary)
- Check for symmetry. (Rise and fall of the chest)

c) Management.

- Ventilate. (Face-mask, bag-valve-mask, if available)
- Alleviate tension pneumothorax.
- Seal open pneumothorax.
- Administer high concentration oxygen, if available.

5) Assess Circulation.

a) Assess and control major bleeding.

- Identify and control major bleeding.
- Pulse: Quality, rate, and regularity.
- Pulse Pressure: Check radial, then carotid.
- Skin color: cyanosis, pallor and temperature.

b) Management.

- Insert two large bore IV's with saline locks. IV fluid replacement, if indicated.
- Place in shock position.
- Prevent hypothermia.

6) Identify priority patients and make transportation decisions.

a) Triage patients at this point.

b) Make transportation decisions. In this type of environment it is possible that a casevac vehicle might not gain access to your position. At this point you will have to organize how the patient will be transported out. Also, you might have to set up relay points and rewarming stations.

C) <u>DETAILED PHYSICAL EXAM</u>: This is the head to toe assessment. We will find and treat all secondary injuries at this time. When performing this exam in the wilderness, you will need to expose one body part at a time and cover it back up as you go. When assessing the patient, you are looking for deformities, contusions, abrasions, penetrations/punctures, bruising, tenderness, lacerations, and swelling (DCAP-BTLS).

1) Head.

a. Assessment:

- Inspect and palpate entire head and face for lacerations, contusions, and fractures.
- Re-evaluate pupils.
- Re-evaluate level of consciousness.
- Assess eyes for retinal hemorrhage, optic disc bulging, visual acuity disturbances, and contact lenses.
- Evaluate cranial nerve function.
- Inspect ears and nose for CSF leakage, hemorrhage.
- Inspect mouth, tongue, and teeth.

b. Management.

- Maintain airway, continue ventilation, and oxygenate as indicated.
- Control hemorrhage.
- Prevent secondary brain injury.
- Remove Contact Lenses.

2) Assess the neck.

a. Assessment.

- Inspect for signs of blunt trauma, penetrating injuries, tracheal deviation, and use of accessory muscles during respiration, and jugular vein distention.
- Palpate for tenderness, deformity, swelling, or subcutaneous emphysema.

b. Management.

- Maintain open airway.
- Maintain immobilization and protection of cervical spine.
- Mark trachea.

3) Assess the chest.

3

a) Assessment.

- Inspect chest wall for signs of blunt trauma, penetrating injuries, and use of accessory muscles during respiration.
- Auscultate the chest wall and posterior bases for breath sounds.
- Auscultate for heart sounds.
- Palpate the entire chest wall for evidence of blunt trauma, penetrating injuries, flail chest, tenderness, crepitus, or subcutaneous emphysema.
- Percuss for hyperresonance or dullness.

c. Management.

- Needle thoracenthesis, if tension pneumothorax is suspected and proper training and equipment are present.
- Three-sided dressing for open chest wounds.
- Pericardiocenthesis, if cardiac tamponade is suspected and proper trained personnel and equipment are present.

4) Abdomen.

a. Assessment.

- Inspect abdomen for signs of blunt trauma, penetrating injuries, and internal bleeding.
- Auscultate for the presence / absence of bowel sounds.

- Palpate the abdomen to elicit subtle tenderness, rebound, or guarding.
- Check pelvis for stability.

b b. Management.

c

- Correctly dress wounds.
- Keep patient NPO until definitive care delivered.

5) Perineum / Rectum / Penis / Vagina.

d a) Assessment.

e

(1) Perineum.

- Contusions / hematomas.
- Lacerations.

(2) Rectum.

- Blood. Sphincter tone.
- High-riding prostate.

(3) Penis.

- Urethral bleeding.

(4) Vagina.

- Bleeding.
- Lacerations.

6) Musculoskeletal.

a) Assessment.

- Inspect extremities for evidence of blunt / penetrating injury. (Deformities, altered range of motion)
- Palpate extremities for tenderness, crepitus, and abnormal movements.
- Palpate all pulses.
- Assess pelvis for stability. (Fractures) - Palpate spine for tenderness, crepitation, step-off's, depressions, and abnormal movements.

b) Management.

- Splint all suspected fractures in the position of function.
- Reduce dislocations as soon as possible, depending on training.
- Maintain immobilization and alignment of cervical, thoracic, and lumbar spine.

7) Reassess the vitals signs.

a) Assessment.

- Assess the vitals, pulse rate, breathing rate, pulse pressure.
- Re-evaluate the pupils and level of consciousness (GCS).
- Evaluate extremities for motor and sensory deficits.

b) Management.

- Maintain adequate immobilization of entire patient, as indicated.

**UNITED STATES MARINE CORPS**
Mountain Warfare Training Center
Bridgeport, California 93517-5001

FMST.07.16
10/22/01

**STUDENT HANDOUT**

**TRIAGE**

**TERMINAL LEARNING OBJECTIVE.** Given multiple casualties in a mountainous environment and necessary equipment and supplies conduct triage in a mountainous environment to prevent death or further injury per the reference. (FMST.07.16)

**ENABLING LEARNING OBJECTIVES.**

1) Without the aid of references, from a given list choose when actual triage most often begins, in accordance with the references. (FMST.07.16a)

2) Without the aid of references, from a given list, identify the four triage categories used by NATO, in accordance with the references. (FMST.07.16b)

3) Without the aid of references, choose the primary cause of death in a mass casualty situation, in accordance with the references. (FMST.07.16c)

**OUTLINE.**

1. **DEFINITION.** Triage or sorting (often used interchangeably) is the evaluation and classification of casualties for the purpose of establishing the priority for treatment and evacuation.

**NOTE**: Casualty sorting is one of the most important tasks for the entire medical service. No other job requires more informed judgment, hard work and courage. The Officer and the Corpsman responsible for triaging wounded Sailors and Marines must exercise sound judgment so that casualties receive proper, expedient care. Triage is a continuing process and any individual assigned should be the most capable and experienced available.

a. Decision Factors. The factors concerning casualty triage are based upon the following:

(1) The need for immediate resuscitation.

(2) The need for emergency surgery.

(3) The futility of surgery due to obviously lethal wounds.

(4) Time needed to perform surgery compared with:

(a) Probability of success.

(b) Number of other casualties needing treatment.

b. The actual triage begins most often with the casualty himself. (FMST.07.16a). This usually occurs when the injured individual determines that "Buddy-Aid" or "Self-Aid" will not be adequate.

(1) The corpsman, once presented with the casualty, must make several determinations:

(a) Is the casualty a walking-wounded or a litter case?

(b) If severely wounded, are resuscitative efforts required immediately?

(c) Does the casualty need any other immediate treatment (bleeding, asphyxia or pain control)?

(2) Once the casualty has been transported to a Battalion Aid Station (BAS), triage continues. The dental officer will commonly perform the duties of the Triage Officer at the echelon I or echelon II level, if present.

2. **NATO TRIAGE CATEGORY CODES**. (FMST.07.16b)

a. At each medical treatment facility (MTF) in the area of operations (AO), incoming casualties are classified by level of treatment required. Both United States and NATO Forces have universally adopted four triage category groups for use. These categories are listed and defined as they appear in standardization agreement (STANAG) No 2879:

(1) Immediate Treatment (Group T1) (RED Tag). To include those requiring emergency life-saving surgery. These procedures should not be time consuming and should concern only those patients with high chances for survival. Some examples are:

(a) Respiratory obstruction.

(b) Accessible hemorrhage.

(c) Emergency amputation.

(2) Delayed Treatment (Group T2) (YELLOW Tag). To include those badly in need of time-consuming major surgery, but whose general condition permits delay in surgical treatment without unduly endangering life. To mitigate the often-critical effects of delay in surgery, sustaining treatment, such as stabilizing I.V. fluids, splinting, administration of antibiotics, catheterization, gastric decompression, and relief of pain will be required. Examples are:

(a) Large muscle wounds.

(b) Fractures of major bones.

(c) Intra-abdominal and/or thoracic, head or spinal injuries.

(d) Uncomplicated major burns.

(3) Minimal Treatment (Group T3) (GREEN Tag). To include those with relatively minor injuries who can effectively care for themselves or who can be helped by untrained personnel. Some examples are:

(a) Minor lacerations.

(b) Abrasions.

(c) Fractures of small bones.

(d) Minor burns.

(4) Expectant Treatment (Group T4) (BLUE Tag). This group comprises patients who have received serious and often multiple injuries, and whose treatment would be time-consuming and complicated with a low chance of survival. If fully treated, they may make heavy demands on medical manpower and supplies. Until the mass casualty situation is under control, they will receive appropriate supportive treatment. The extent of treatment depends on available supplies and manpower, and may involve the use of large doses of narcotic analgesics. These patients should not be abandoned, and every effort should be devoted to their comfort. The possibility of their survival, despite alarming injuries, must always be kept in mind. Examples are:

(a) Severe multiple injuries.

(b) Severe head or spinal injuries.

(c) Large doses of radiation.

(d) Widespread severe burns.

3. **TREATMENT CONCEPTS**.

a. The triage officer must be familiar with all aspects of combat casualty care (airway management, hemorrhage control, use of anesthesia, surgical procedures, fluid resuscitation and post-operative care, to name a few), but ever-present in his mind should be the following concepts:

(1) Asphyxia and hemorrhage are initially the primary causes of death in any mass casualty situation. (FMST.07.16c)

(2) Once at a definitive care facility, uncontrolled hemorrhage is the leading cause of death.

(3) Life has priority over limb, and preserving function overrides the correction of superficial anatomic defects.

(4) A casualty is in a constant state of change until the wound or injury has been repaired.

(5) Systemic disturbances caused by the wound continue until healing is complete.

(6) Combat conditions will dictate the modifications necessary in triage and care of mass casualties.

b. Priorities of Treatment. With the six concepts of treatment in mind, we can now address three basic priorities for treatment.

(1) First Priority:

(a) Asphyxia:

1. Mechanical obstruction of the airway.

2. Sucking chest wound.

3. Tension pneumothorax.

4. Maxillofacial trauma.

5. Massive hemo-pneumothorax.

(b) Shock:

1. Exsanguinating hemorrhage.

2. Major internal hemorrhage (thoracic, abdominal, and pelvic).

3. Visceral injuries/evisceration.

4. Cardio-pericardial injuries.

5. Massive muscle injury leading to Fat Embolization to the lung.

6. Full-thickness or deep-partial burns of 10 % or greater.

(2) Second Priority:

(a) Visceral injuries:

1. Perforation of G.I. Tract.

2. Pancreatic and Biliary system.

3. Genitourinary tract.

4. Thoracic wounds w/o asphyxia.

(b) Vascular injuries requiring repair:

1. Tourniquet cases.

(c) Closed head injuries:

1. Increasing loss of consciousness.

(d) Major burns:

1. Partial-thickness burns of greater than 25% TBSA in the low risk group, 20% TBSA in the high-risk group.

2. Burns involving the poor risk group.

3. Burns involving the face, hands, feet, genitalia/perineum.

4. Circumferential limb burns.

(3) Third Priority:

(a) Brain and spinal injuries - requiring burr hole decompression.

(b) Soft tissue injuries.

(c) Debridement necessary with less than major muscle damage.

(d) Lesser fractures and dislocations.

(e) Eye injuries.

(f) Maxillofacial injuries w/o asphyxia.

(g) Burns under 20% in other locations other than priority 2.

4. **TRIAGE TOOLS**.

a. Trauma Score/Glasgow Coma Scale. In recent years, many methods have been proposed to speed up the classification and sorting of casualties. One method that has gained wide acceptance is the Trauma Score/Glasgow Coma Scale. Several studies have been performed using various parameters, i.e., respiratory rate, chest wall expansion, BP, capillary refill, etc. By assigning a number scale to each specific response, the likelihood of survival as a percentage can be calculated.

1. Glasgow Coma Scale.

a. Eye opening:

1) Spontaneous = 4.
2) To voice = 3.
3) To pain = 2.
4) None = 1.

b. Verbal Response.

1) Oriented = 5.
2) Confused = 4.
3) Inappropriate words = 3.
4) Incomprehensive words = 2.
5) None = 1.

c. Motor Response.

1) Obeys Command = 6.
2) Localized Pain = 5.

        3) Withdraw Pain = 4.
        4) Flexion (pain) = 3.
        5) Extension (pain) = 2.
        6) None = 1.

    d. Total the points up and they equal:

        1) 13-15= 4.
        2) 9-12 = 3.
        3) 6-8 = 2.
        4) 4-5 = 1.
        5) <3 = 0.

2. Trauma Scale.

    a. Respiratory Rate:

        1) 12-29 min = 4.
        2) 10-12 min = 3.
        3) 6-9 min = 2.
        4) 1-5 min = 1.
        5) None = 0.

    b. Systolic BP:

        1) >89 mmHg = 4.
        2) 76-89 mmHg = 3.
        3) 50-75 mmHg = 2.
        4) 1-49 mmHg = 1.
        5) None = 0.

    c. Total the points.

3. Total trauma Scale + GCS = % survival (1-12):

    12 = 99.
    10 = 88.
    8 = 67.
    6 = 63.
    4 = 33.
    2 = 29.
    1 = 25.
    0 = 4.

**UNITED STATES MARINE CORPS**
Mountain Warfare Training Center
Bridgeport, California 93517-5001

FMST.07.15
10/22/01

**STUDENT HANDOUT**

# COMBAT CASUALTY CARE

**TERMINAL LEARNING OBJECTIVE.** Given a casualty in a mountainous environment and necessary equipment and supplies, perform combat casualty care procedures, in accordance with the references. (FMST.07.15)

**ENABLING LEARNING OBJECTIVE.**

(1)Without the aid of references, choose from a given list the three correct phases of tactical casualty management, in accordance with the references. (FMST.07.15a)

(2) Without the aid of references, given a simulated casualty and the necessary equipment, conduct combat casualty care, in accordance with the references. (FMST.07.15b)

**OUTLINE**

1. **CONCEPT:** The most important aspect of caring for trauma victims on the battlefield is well thought out planning for that environment and appropriate training of combat medical personnel.

2. **Phases of Care.** (FMST.07.15a) In making the transition from the standards of ATLS to the tactical setting, it is useful to consider the management of casualties that occur during tactical missions as being divided into three distinct phases:

   a. Care Under Fire. Care rendered at the scene of the injury, while still under effective hostile fire.

   b. Tactical Field Care. Care rendered when no longer under effective hostile fire or when there has been no hostile fire.

   c. Combat Casualty Evacuation. Care rendered once the casualty has been picked up by an aircraft, naval craft or vehicle.

**10**

3. **Phase One: Care Under Fire**

a. Basic Casualty Management Plan.

1. Return of fire as directed.

   a. The risk of injury to other mission personnel and additional injury to previously injured operators will be reduced if immediate attention is directed to the suppression of hostile fire.

   b. The best medicine on any battlefield is fire superiority.

2. Try to keep yourself from getting shot.

   a. There are typically one or two corpsmen assigned to each small operating unit. If they sustain injuries, the management of casualties in this phase is drastically reduced.

3. Try to keep casualty from sustaining additional wounds (the first priority once you are able to render care).

   a. Operators should be taught, if they are wounded and unable to participate further in the engagement, they should lay motionless on the ground, if covered, or to move as quickly as possible to nearby cover. If no cover is available, they should remain motionless on the ground so as not to draw more fire.

4. Stop any life threatening external hemorrhage with a tourniquet.

   a. Injuries to major vessels can quickly lead to hypovolemic shock.

   b. If the casualty needs to be moved, as is usually the case, a tourniquet is the most reliable initial choice to stop bleeding.

5. Take the casualty with you when you leave.

   a. Transport of the patient is currently accomplished with an improvised method of carry.

      1. While under fire, hasty methods are used such as the clothes drag, shoulder drag, shoulder carry or the extremity carry.

      2. When fire superiority is accomplished, improvised litters may be used due to the decreased risks of exposure to uninjured operators.

4. **Phase Two: Tactical Field Phase**

   a. This phase is distinguished from the care under fire phase by more time with which to render care and a reduced level of hazard from hostile fire.

b. Time to render care will be variable and situation dependent.

c. Basic Tactical Casualty Management Plan for Phase Two. As taught in ATLS, attention is first directed to the evaluation of airway, breathing and circulation.

1. Airway Management.

   a. Chin-lift or Jaw-thrust.

   b. Unconscious casualty without airway obstruction: Nasopharyngeal airway.

   c. Unconscious casualty with airway obstruction : Cricothyroidotomy.

   d. Cervical spine immobilization is not necessary for casualties with penetrating head or neck trauma.

2. Breathing.

   a. Consider tension pneumothorax and decompress with needle thoracostomy if a casualty has unilateral penetrating chest trauma and progressive respiratory distress.

3. Bleeding.

   a. Control any remaining bleeding with a tourniquet or direct pressure.

4. IV.

   a. Start an 18-guage IV or saline lock.

5. Fluid Resuscitation.

   a. Controlled hemorrhage without shock:
   No fluids necessary.

   b. Controlled hemorrhage with shock:
   Hespan 1,000 cc.

   c. Uncontrolled (intra-abdominal or thoracic) hemorrhage:
   No IV fluid resuscitation .

6. Inspect and dress wound.

7. Check for additional wounds.

8. Analgesia as necessary.

a. Morphine: 5 mg IV, wait 10 minutes; repeat as necessary.

9. Splint fractures and recheck pulse.

10. Antibiotics.

a. Cefoxitin: 2g slow IV push (over 3-5 minutes) for penetrating abdominal trauma, massive soft tissue damage, open fractures, grossly contaminated wounds or long delays before casualty evacuation.

11. Cardiopulmonary Resuscitation.

a. Resuscitation on the battlefield for victims of blast or penetrating trauma who have no pulse, no respirations, and no other signs of life will not be successful and should not be attempted.

5. **Phase Three: Combat Casualty Evacuation**

a. At some point in the operation, the mission personnel will be recovered onto an aircraft, naval craft or other asset for extraction.

b. Time to extraction can be quite variable.

c c. Generally, other mission personnel will be extracted with casualties, depending on mission requirements.

d. MEDEVAC vs CASEVAC

1. The term "Medevac" is reserved for the aeromedical evacuation of a stable patient from one medical treatment facility to another.

2. Thus the term "Medevac" should be avoided when discussing the initial management of combat casualties and the term "Combat Casualty Evacuation" or "Casevac" be used instead to eliminate any misunderstanding of the mission required.

e. There are two significant differences in the progression from the tactical field care phase
d to the Casevac phase:

1. Additional medical personnel may accompany the evacuating asset. The possibility of having more highly trained and experienced medical personnel at this point of the operation should not go to waste.

2. Additional medical equipment may be pre-staged on the evacuating asset.

3. Coordination for these additional medical assets should be arranged during the mission preparation phase.

f. <u>Basic Tactical Casualty Management Plan for Phase Three</u>:

1. <u>Airway.</u>

   a. Chin-lift, Jaw-thrust.

   b. Unconscious casualty without airway obstruction: Nasopharyngeal airway, endotracheal intubation, Combitube or laryngeal mask airway.

   c. Unconscious casualty with airway obstruction: Cricothyroidotomy if endotracheal intubation and / or other airway devices are unsuccessful.

2. <u>Breathing.</u>

   a. Consider tension pneumothorax and decompress with needle thoracostomy if a casualty has unilateral penetrating chest trauma and progressive respiratory distress.

   b. Consider chest tube insertion if a suspected tension pneumothorax is not relieved by needle thoracostomy.

   c. Oxygen.

3. <u>Bleeding.</u>

   a. Consider removing tourniquets and using direct pressure to control bleeding.

4. <u>IV.</u>

   a. Start an 18-guage IV or saline lock if not already done.

5. <u>Fluid Resuscitation.</u>

   a. No hemorrhage or controlled hemorrhage without shock: Lactated Ringer's at 250 cc / hour.

   b. Controlled hemorrhage with shock: Hespan 1,000 cc initially.

   c. Uncontrolled (intra-abdominal or thoracic) hemorrhage: No IV fluid resuscitation.

   d. Head wound patient: Use colloids instead of crystalloids to decrease chance of raising intra-cranial pressure.

e. Hespan at a minimal flow to maintain infusion unless there is concurrent controlled hemorrhagic shock.

6. Monitoring.

a. Institute electronic monitoring of heart rate, blood pressure and hemoglobin oxygen saturation (pulse oximetry).

7. Inspect and dress wound if not already done.

8. Check for additional wounds.
8
9. Analgesia.

a. Morphine: 5 mg IV; wait 10 minutes; repeat as necessary.

10. Splint fractures and recheck pulse if not already done.

11. Antibiotics.

a. Cefoxitin: 2 g slow IV push (over 3-5 minutes) for penetrating adbominal trauma, massive soft-tissue damage, open fractures, grossly contaminated wounds or long delays before reaching a definitive medical treatment facility.

**UNITED STATES MARINE CORPS**
Mountain Warfare Training
Center Bridgeport, California
93517-5001

FMST.07.22
10/23/01

## STUDENT HANDOUT

## LAND NAVIGATION REVIEW

**TERMINAL LEARNING OBJECTIVE.** In a mountainous environment, navigate in mountainous terrain, in accordance with the references. (FMST.07.22)

**ENABLING LEARNING OBJECTIVES**

1) Given a map and compass, orientate the map, in accordance with the references. (FMST.07.22a)

2) Given a map and a protractor locate a six-digit grid coordinate, in accordance with the references. (FMST.07.22b)

3) Given a map, compass, and protractor perform a resection to locate an unknown point from three known points, in accordance with the references. (FMST.07.22c)

4) Without the aid of references, select from a given list the correct definition for true north, grid north, and magnetic north, in accordance with the references. (FMST.07.22d)

**OUTLINE.**

1. **THE MAP**. Before you can properly use a map, there is some basic information you will need to understand.

a. Definition. A map is a reduced or scale drawing of the ground and important things on the ground as seen from the air. Essentially it is a picture of the surface of the earth, as it would appear looking at it from an aircraft.

b. Marginal Information. As you take a quick look at the borders, you will notice that there is a large amount of information. When using a map, the instructions are placed around the outer edges of the map and are known as marginal information. All maps are not the same, some having more marginal information than others, and the information being located in different places around the border. It is important that you know how to use this information if you plan to get the most out of your map. The different information on the margins is as follows:

(1) Sheet name. The sheet name is found in two places: the center of the upper margin and lower right margin of the map. The map is usually named after its outstanding cultural or geographical features. Whenever possible, the name of the largest city is used.

(2) The sheet number. The sheet number is found in the upper right margin of your map and is used as a reference number for the map sheet.

(3) Series name and scale.

(a) The series name is found in the upper left margin of your map. A map series usually comprises a group of similar maps on the same scale and on the same sheet lines or format designed to cover a particular geographical area. It may also be a group of maps designed to serve a common purpose. The name given a series is of the most prominent area.

(b) The scale notation represents the ratio of the maps distance to the corresponding (actual) distance on the earth's surface. For example, the scale notation of 1:50,000 indicate that one unit of measure on the map equals 50,000 of the same units of measure on the actual surface of the ground (1 map inch = 50,000 earth inches).

(4) Series number. The series number appears in the upper right margin and lower left margin. It is a comprehensive reference expressed either as a four digit number or as a letter followed by three or four numbers.

(5) Index to adjoining sheets. The index to adjoining sheets is found in the lower right margin of your map.

(6) Bar scales. The bar scale is located in the center lower margin of your map.

(7) Contour interval. The contour interval is located in the center lower margin of your map, just below the bar scales. It states the vertical distance between adjacent contour lines on the map. For example, every contour line represents an increase in elevation as indicated by the distance of the contour interval.

(8) Declination diagram. The declination diagram is located in the right portion of the lower margin of your map.

(a) True North. (FMST.07.22d) A line from any position on the earth's surface to the North Pole, which is the exact top of the earth.

(b) Grid North. The north direction established by the vertical grid lines on the map.

(c) Magnetic North. The direction to the North Magnetic Pole.

(9) Legend. Is located in the lower left margin. It illustrates and identifies some of the symbols on the map. There is not enough room on a map sheet to show the true outlines of objects and land features so the mapmaker uses a set of standard symbols to represent them. Sometimes the symbols are not the same or standard types of symbols typically used. The type of map, the scale, or the origin causes this. Because of this possible difference, to prevent errors in symbol identification, the legend should be referred to every time a map is used. (The example on the legend may state, "unimproved road, hard surface road, jeep trail," etc. The right side of the symbol will have three symbols; the first being "unimproved road", the second "hard surface road", and the third being "jeep trail".)

(10) Map colors. The colors you see on your map are not there to make the map pretty or enjoyable to the user. They have important meaning. The colors vary with different types of maps, but on a standard, large scale map, there are five basic colors:

(a) Black. Represents most of the man-made features.

(b) Red. Represents major roads.

(c) Blue. Represents water.

(d) Green. Represents vegetation.

(e) Brown. Represents all terrain features, such as hills, draws, fingers, depressions, and saddles by the use of contour lines. These contour lines are shown in brown.

(f) Other Colors. Occasionally other colors are used to show special information. These, as a rule, will be indicated in the marginal information. Currently, there are changes being made to the color system, which may result in changes of the marginal information in regards to map colors.

2. **HORIZONTAL SCALES**.

a. Bar Scale. This special ruler, located at the bottom of the map sheet, enables you to measure a distance between 2 points on your map and change that map distance into the actual ground distance.

(1) The primary scale. The primary scale is the part of the scale that starts at the zero (0) and extends to the right.

(2) The extension scale. The second part of the bar scale is the extension scale. It starts at the zero (0) and extends left.

b. Measuring Straight-Line Distance. If you are given a mission that requires you to move from your present position to a different position some distance away and you plan to travel in a straight line between these two points, the following method is used to determine the distance you will have to travel:

(1) First locate your present position and your destination on your map. Then you will lay a straight edged piece of paper on the map so that the edge of the paper touches the location of your present position and the location of your destination. Make a tick mark on the edge of the paper at each point. Now move the paper down to the bar scale. Place the left tick mark on the zero (0) of the scale so that the other tick mark will fall somewhere on the primary scale.

c. Measuring a Curved Line Distance. If you are given a mission that requires you to move from your present position to a different position and this time you choose to move along a road to reach your destination, the following method is used to work out the ground distance you will have to travel:

(1) Find your present position, your destination, and the road you will follow, on your map. Position a straight edged piece of paper at your location and make a tick mark. Now, rotate the paper so that it runs along the road until the road turns, and make a tick mark on both the paper and the map. Keeping these 2 tick marks together, rotate the paper again until the road turns, and again mark the paper and the map. Continue this procedure until you have reached your destination. Now using you're bar scale as previously stated, find the actual ground distance your squad will be traveling.

3. **ELEVATION AND RELIEF**. In order to begin learning the subject of elevation and relief, let's first look at some definitions.

a. Contour lines. Of the several ways to indicate elevation and relief on maps, the most common is through the use of contour lines. A contour line represents an imaginary line on the ground along which all points on this line are the same elevation above sea level.

b. Characteristics of Contour Lines. There are a few characteristics of a contour line that you should know that will help you understand what they represent on your map. These characteristics are as follows:

(1) Contour lines never cross.

(2) Contour lines connect on themselves.

c. Types of Contour Lines. In addition to knowing the definition, purpose, and characteristics of a contour line, you need to realize that there are different types of contour lines.

(1) Index contour lines. Starting at zero elevation, every 5th contour line is drawn with a heavier line.

(2) Intermediate contour lines. These are the contour lines falling between the Index contour lines.

(3) Supplementary contour lines. These lines are shown as dashed brown lines.

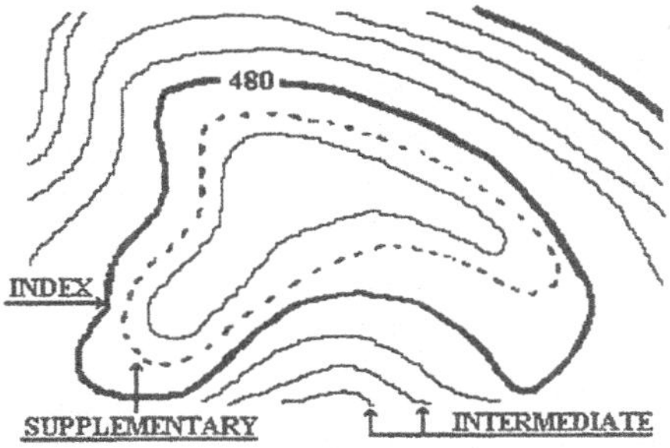


d. Determining Elevation of a Point. Now you are ready to learn how to determine the elevation of any point of the ground represented on your map.

(1) Given elevation nearest the point, examine the area around your point until you find the nearest point that has a marked elevation. There are a few different ways that contour lines are marked, and they are:

(a) Labeled index contour line. Index contour lines are broken with the elevation of the line printed in the break.

(b) Spot elevations. At various places on your map, usually on prominent landforms such as road junctions or hilltops, a spot elevation will be given. It will be printed in either brown or black.

(c) Bench marks. Similar to Spot elevations, but more accurate. They are marked on your map by a black "x", with the letters "BM", and the elevation.

(d) A point on a contour line. If a point is located on a contour line, its elevation will be the same as that contour line.

(e) A point between contour lines. If a desired point is less than 1/4 the distance between 2 lines, its elevation is considered to be the same as the closest contour line. If the point is from 1/4 to 3/4 the distance, its elevation is considered to be in the middle of the 2 lines.

(f) A point on top of a hill. To estimate the elevation on the top of an unmarked hill, add 1/2 of the contour interval to the elevation of the highest contour line around the hill.

4. **TERRAIN FEATURES**. There are many types of terrain features that you must be able to identify. By interpreting the arrangement of contour lines, you should be able to get a

mental picture of how these terrain features would appear if you were actually out there on the ground looking at them. Some of the features you need to be able to recognize are as follows:

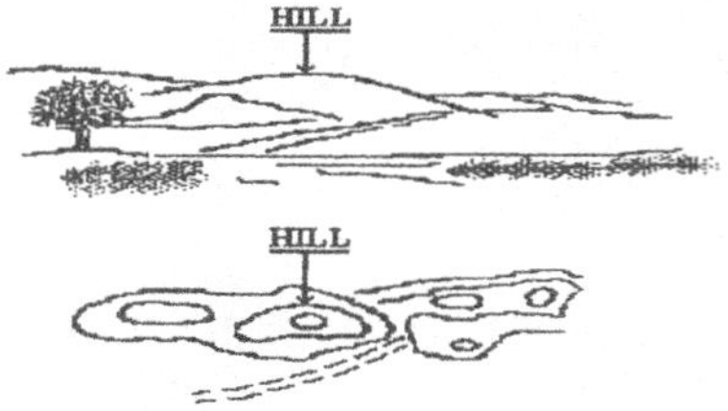


a. Hills. Defined as a point or small area of high ground. A hill is represented on your map by a number of contour lines each of which circles around and connects on it. These rough circles get smaller as they progress toward the top of the hill. You need to carefully compare the arrangement of these contour lines to the actual appearance of the hill.

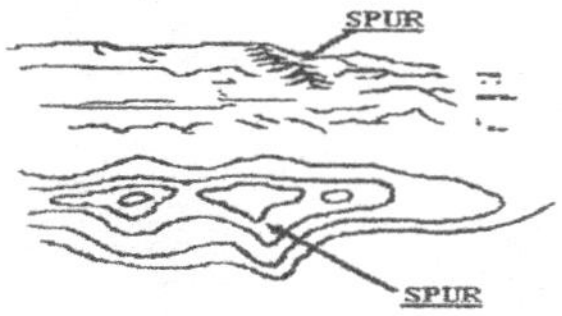


b. Spur. Defined as a line of high ground. The points along the top or crest of the spur are higher than the ground on both sides. A spur is represented on the map by a number of contour lines, each of which form an U-Shaped or V-Shaped design. The end or bottom of the U or V-Shaped lines point down to a lower elevation.

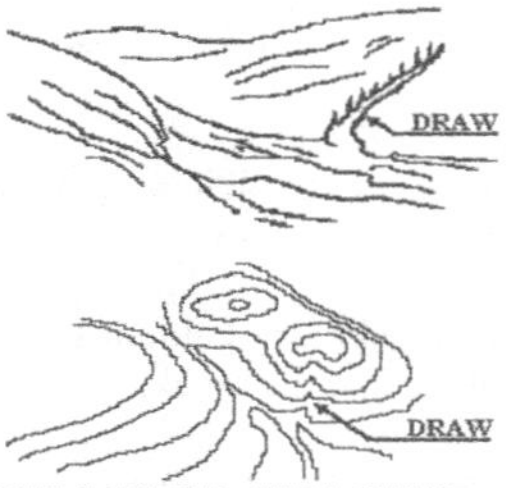


c. Draw. Defined as a line of low ground. The points located in the bottom of a draw are at a lower elevation than the ground to either side. A draw usually runs between 2 spurs. It is sometimes difficult to tell the difference between the contour lines that represent a draw from those that represent a spur because they are EITHER U or V-Shaped lines. Remember that the curved portion representing a spur points down to ground of lower elevation, and the curved portion representing a draw points to ground of higher elevation. Now by looking closely at the area on your map, which of the 2 you are dealing with. In many cases, draws will contain a stream, while spurs usually never have a stream running down their centers.

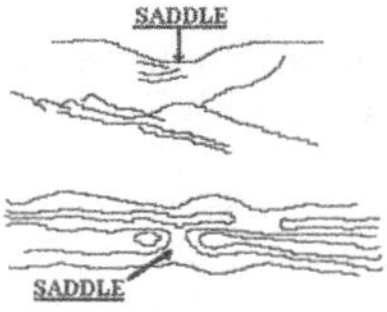


d. Saddle. Defined as the low ground between 2 hilltops, or a dip along the crest of a spur. On your map a saddle is represented by contour lines that form 2 hilltops that are located closely to each other. The area between these 2 hilltops being the location of the saddle.

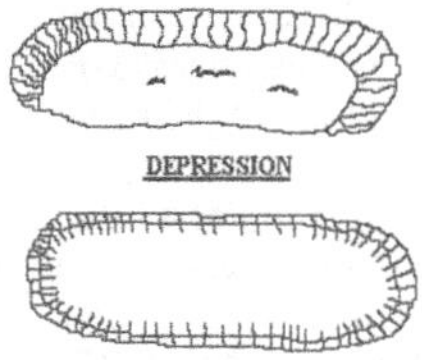


e. Depression. Defined as a point or area of low ground surrounded on all sides by higher ground. A depression is represented by contour lines forming circles within a small area on the map. These rough circles look identical to hilltops except for the presence of hachures. These are short stubby lines that are connected to the contour line at one end and point away from the contour lines at right angles in a downhill direction.

f. Cut. A cut is a man-made feature by which the bed of a road or railroad is leveled by cutting through high ground. A cut is represented by a sudden ending of contour lines, the ends being connected by a straight line. Hachures are connected to this straight line and point downhill toward the roadbed. This sudden ending of contour lines occurs on both sides of the roadbed.

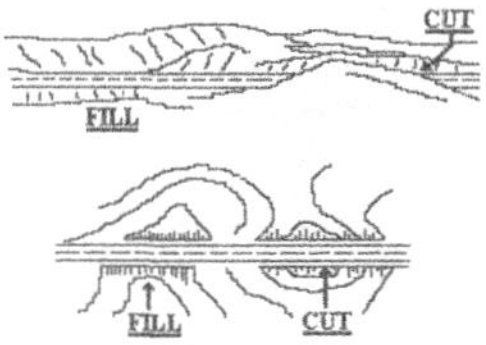


g. Fill. A fill is also a man-made feature by which the bed of a road or railroad is leveled by filling in the low area. A sudden ending of contour lines represents a fill also. A straight line connects the ends with Hachures pointing downhill away from the roadbed. Again, this occurs on both sides of the roadbed.

5. **SLOPES.** The rate of rise or fall of a ground form is known as its slope. In combat, this factor of slope must be considered when making plans for an operation. You must have the ability to recognize the ground forms in your area of operation by studying your map and the spacing between the contour lines on your map.

CONCAVE SLOPE

CONVEX SLOPE

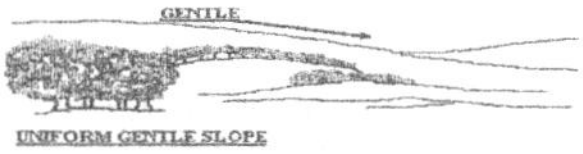


UNIFORM GENTLE SLOPE

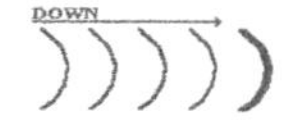


CONVEX SLOPE

scribe tl
neasure

(1) Grid north. It is used when measuring direction from your map.

(2) Magnetic north. Used when measuring direction from your compass.

b. Units of Measurement. Now that you have learned about the two base direction lines that are used to measure direction, we need to learn the units of measurement that are used to measure the angle from one of these base direction lines to the actual direction line.

(1) Degrees. The degree is the unit of measurement that is used most often. Anytime you need an azimuth to move from one point to another, you will use degrees as your unit of measurement. The base direction line, grid north or magnetic north, will be zero (0) degrees and measuring clockwise from this base direction line, the angle increases until you reach 360 degrees which brings you back around in a full circle to the base direction line.

(2) Mils. The mil is also used as a unit of measurement for direction. It is used when working with indirect fire weapons such as mortars and artillery. The azimuth you send will have to

be in mils, not in degrees. Remember, you are still measuring the angle between the base direction line and the line of sight from you to the distant object only you are using a different unit to measure the angle. Moving clockwise from the base direction line, which would be zero (0) mils, the angle back to the direction line. Now you know there are 360 degrees and also 6400 mils in the circle. This means that each degree equals approximately 17.5 mils.

c. Converting Map and Compass Directions. (CWM.6.22d) Earlier you learned that an azimuth was always measured from a base direction line. You also learned that we could use two different base direction lines. Grid north is the base direction line used when you are working with the map and getting a map direction or a grid azimuth. Magnetic north is the base direction line used when you are using the compass to get a compass direction or a magnetic azimuth. Because these two different base direction lines are used and because in most cases there is a difference of one or more degrees between them, it is necessary for you to learn how to convert from a grid azimuth to a magnetic azimuth or from a magnetic azimuth to a grid azimuth. In order to learn this procedure we must take a look at the declination diagram.

(1) Declination diagram. A diagram that shows the relation between three different norths: true north, grid north and magnetic north. It is the difference between grid north and magnetic north that we need to look at. The difference between these two norths is known as the grid-magnetic angle or G-M angle.

(a) Grid-Magnetic angle. The grid-magnetic angle is the angular difference between grid north and magnetic north. As you can see on this diagram a line that connects grid north and magnetic north represents the G-M angle. The size of the angle is stated off to the left side of the diagram in both degrees and mils. The G-M angle on your maps is 18 degrees or 320 mils. It is important to know that the G-M angle will change depending on where you are located on the earth's surface.

(2) Converting. When you receive a mission as a squad leader to move your squad from your present position to some distant point, you take out your map, a protractor and figure out the map direction or the grid azimuth to that distant point. However, before you can use your compass to direct you to your destination, you will have to convert the grid azimuth you have from your map, to a magnetic azimuth that you can set on your compass.

(a) The first step in making this conversion is to examine the declination diagram on your map. The G-M angle on your map is 18 degrees. You know that you will have to add the G-M angle to, or subtract it from the azimuth you measured on your map, or shot on your compass.

(b) The instructions as to whether you will add or subtract this angle are written on the map depending on whether you are going from grid to magnetic or vise-versa.

d. Orientation of a Map. (CWM.6.22a) In order to gain the best use of your map and to avoid confusion while navigating across terrain, you will need to know how to orient your map to the ground. A map is oriented when it is lying flat and level with it's north and south pointing in the same directions as north and south on the ground. There are two methods you can use to orient a map.

(1) Inspection. The first method used to orient the map is by inspection. By carefully examining your map and the ground around the area in which you are located to find linear features such as roads, railroads and power line, you can align the feature on the map with the same feature on the ground. If there is only one linear terrain feature in the area in which you are located, you will have to be careful not to reverse your directions. Even though you have a road running north south on your map aligned with the same road on the ground, the north end of your map may be pointed in a southerly direction down the road. To avoid this from occurring, find another terrain feature in the area that you can also identify on the map. Once you have aligned your map along the linear feature, check to insure that the second feature is positioned on the same side of the linear feature on your map as it is on the ground.

(2) Compass. This is the second method of orienting a map. You begin by placing your map on the ground so it lies flat and level. Open your compass so that the cover and eyepiece are completely open. Align the graduated straight edge of the compass along one of the north-south grid lines on your map with the cover end of the compass pointing toward the top of the map. This will place the stationary index line parallel to the grid north. Since the north-seeking arrow on the dial of your compass will always point to magnetic north, you set your compass up just like a declination diagram. Grid north being the stationary index line and magnetic north being your north-seeking arrow. Now rotate your map, insuring that the graduated straight edge of your compass remains aligned with the north-south grid line, until the face of the compass appears exactly as the declination diagram in the lower margin of the map. Your map is now oriented. For the Sonora Pass map sheet you would rotate the map until the north-seeking arrow on the compass lies 18 degrees to the right of the stationary index line.

7. **GRID SYSTEM.** (CWM.6.22b) In most combat situations it is very important to be able to pinpoint your location, the location of another unit, or the location of an objective. In order to do this, it is necessary to have a location system that will allow you to accurately determine a specific location and state that location in such a way that will allow others to find that location quickly. This system is known as a grid system.

a. Grid Box. This box gives basic instructions on reading grids in determination of specific points on the map.

b. Grid Lines. The straight black lines that run across, and up and down the surface of your maps form the 1000-meter grid squares. These lines are called grid lines. If you look around the border of your map you will see that a two-digit number identifies both the horizontal and vertical grid lines.

c. Grid Square Identification. Each 1000-meter grid square on your map has its own unique number. This number is called a grid coordinate. A four digit number is the result of combining the two digit number identification of the vertical grid line and the two digit number identification of the horizontal grid line that cross to form the lower left comer of the grid square. To find the grid coordinate of any grid square on your map you will always read your map right and up. You find the two-digit number of that vertical grid line and you will have the first 2 digits of your coordinate. Next, find the two-digit number of the horizontal grid line and place it after your vertical number and you now have the 4-digit grid coordinate that identifies your grid square.

d. Locating Terrain Features Within a Grid Square. If you are given a 4 digit grid coordinate to find a certain point on the earth's surface, that grid coordinate would put you within 1000 meters of that point. If you are told to find a certain feature within that grid square and you are good at reading your map, a 4-digit coordinate may get you close enough to your point to enable you to find it. However, there are many situations in which a 4-digit number will not get you close enough to a given position to allow you to accomplish your task. It is necessary to be able to locate points within a grid square and using a six-digit grid coordinate does this.

(1) Six Digit Grid Coordinate. The 1000-meter grid square will have to be broken down further into 100-meter grid squares. Because the surface of your map would become too cluttered by all these additional lines, a special instrument has been made to help you divide up a 1000-meter grid square. This instrument is called a coordinate scale or protractor.

8. **RESECTION**. (CWM.6.22c)

a. You may face a situation where you need to pinpoint the location of your position. This may not be possible by associating the terrain in your area with the map because you may have a poor map or the terrain in your immediate area is flat and has no distinguishing features. When you find yourself in this situation you will need to know how to perform a resection.

(1) Examine the terrain around your position and find two terrain features or objects that you can also identify on your map.

(2) Use your compass to find the magnetic azimuth to one of the distant terrain features you have identified.

(3) Convert the magnetic azimuth you found into a grid azimuth.

(4) Change the grid azimuth you found into a back azimuth.

(5) Drawing a line on your map from a distant point you identified in step one back toward your present position. Place a small mark on your map at this point. Using a straight edge, draw a line from where you plotted the distant object to the mark you made by the edge of your protractor. This line should be extended to insure it continues through the area where you are located.

(6) Repeat, finding the magnetic azimuth to the second distant object. That is at least 60 degrees from the first.

(7) When you have drawn your second line on the map, the point where the second line crosses the first line is the location of your position. The process that has just been covered is called a two-point resection.

(8) If your position is located on a linear line feature such as a road, powerline, railroad or river, then it is only necessary to perform an one-point resection in order to pinpoint your location. Instead of identifying two distant points, you will only need to select one. Where the single line that you drew on your map crosses the linear feature on which you are located will be the location of your position.

9. **DETOURING AROUND AN OBSTACLE**.

a. There will be occasions when it will become necessary to move off your plotted route in order to avoid an obstacle such as a small lake or swamp. You must be able to move around these obstacles and return to the route you were following before you took your detour. This can be accomplished by using one of three detour methods: Farside landmark method, nearside landmark method, and the 90-degree offset method.

(1) Farside landmark. Used during hours of daylight when you are able to identify some feature on the farside of the obstacle that lies on your line of March and is obvious enough so that you can find it once you have moved around to the farside.

(2) Nearside landmark. Used during hours of daylight when you cannot find a landmark on the farside. However, you can identify a feature on the nearside that is on your. Line of March and you will still be able to see once you have moved around to the farside. Again, let us use the example of being on an azimuth of 198 degrees, and it brings you to a small lake. From the nearside you cannot locate any features on the farside. There is, however, a small pier on the nearside that is located on your line of March. You can now move around the lake using the easiest and safest route. Move to an area on the farside that is as close to being opposite of the nearside landmark as possible. You then figure out the back azimuth of your route azimuth, which in this case would be 18 degrees. Shoot an azimuth back across the lake to the pier. Move to the right or left until you are in a position where the pier lies at an azimuth of 18 degrees. This puts you back on your line of March and you can continue on following your original azimuth of 198 degrees.

(3) 90 degree offset. Used during hours of reduced visibility or during daylight hours when there are no visible landmarks along the line of March on either side of the obstacle. This time you are moving along 360 degrees azimuth and you encounter a large, impassable area of thick brush. The brush is too high to see any landmarks from either side of the thicket. The following procedure should be taken to detour around the thicket.

(a) Stop the pace count you have been making for movement along your assigned route.

(b) Decide which way you will detour around the thicket. Let's say in this case the right flank appears to be the easiest and safest. Add 90 degrees to your original azimuth of 360 degrees in order to move your patrol to the right.

(c) Move on this 90 degrees azimuth until you have reached a point where you can clear the right flank of the thicket. You will keep a new pace count starting from the point where you made your 90 degrees turn to the right to the point where you can clear the right flank of the thicket.

(d) Make another change of direction. This time you want to return to a 360 degrees azimuth so you will parallel your original course and move by the right flank of the thicket. In order to return to the 360 degrees azimuth you will now subtract 90 degrees.

(e) Move on the 360-degree azimuth until you have by passed the thicket. At the point of your second 90-degree turn, you will pick up the pace count you were making before you began making your detour around the obstacle. Even though you are not on your original line of march, you are moving parallel to your original line of march and therefore must include the distance you cover during this leg of the offset in your pace count for your route.

(f) Once you have bypassed the thicket, you will make another change of direction. This time, you want to move back toward your original line of March. Since that means you will be turning toward the left, you will subtract 90 degrees from the 360 degrees azimuth you have just completed. Your new direction will be 270 degrees.

(g) Pace back toward the original line of March following the 270-degree azimuth, the same number of paces as you made off the original line of March. In doing this you will have returned to your original line of March.

(h) Once you have moved back to your original line of March you will make your last change of direction. By adding 90 degrees to the original azimuth of the last leg of the offset, which in this case was 270 degrees, you will be back on the original azimuth of 360 degrees.

(i) Return to your original line of March pace count.

**UNITED STATES MARINE CORPS**
Mountain Warfare Training Center
Bridgeport, California 93517-5001

FMST.07.01
10/23/01

**STUDENT HANDOUT**

**NUTRITION**

**TERMINAL LEARNING OBJECTIVE** Given a unit in a mountainous environment, and the necessary equipment and supplies, apply the principles of nutrition in a mountainous environment to prevent death or injury per the reference. (FMST.07.01)

**ENABLING LEARNING OBJECTIVES**

1) Without the aid of references and from a given list, choose the correct caloric intake needed per day in a mountainous environment, in accordance with the references. (FMST.07.01a)

2) Without the aid of references and from a given list, choose which type of diet increases your tolerance to cold, in accordance with the references. (FMST.07.01b)

3) Without the aid of references and from a given list, choose one of the preventive measures of excessive water loss in a mountainous terrain, in accordance with the references. (FMST.07.01c)

**OUTLINE.**

1. **STARVATION FACTORS.** Factors that combine to result in "Near Starvation" during military operations in any mountainous environment:

a. Difficulty in carrying sufficient food.

b. Lack of time for preparing meals and beverages.

c. Loss of appetite and low food palatability resulting from hypoxia/hypothermia.

d. Fat intolerance resulting from altered metabolism.

2. **PHYSIOLOGICAL ASPECT TO COLD STRESS.**

a. Man's adaptation to cold is essentially behavioral. Shelters and clothing are utilized to provide a micro-environment, which will allow survival. Body composition adjustments such as increases in body weight and fat are also examples of the body's adaptation to the cold. The winter months lend themselves to greater periods of inactivity, leading to these changes.

b. In the Falkland War, constipation was a major concern. Dehydration was found to be the culprit in the majority of cases, but diet had a significant role as well. MRE's are low in fiber, containing less than 10 grams. Man needs at least 31 grams of fiber per day to ease passage of stool. Constipation can be a debilitating condition, affecting overall unit readiness.

(1) Low fiber diets can lead to diverticulosis, diverticulitis and Irritable Bowel Syndrome (IBS). IBS flare-ups tend to occur during upsetting situations and emotional stress. Educate your troops on the importance of fiber, especially in a cold weather environment.

(2) Fiber is supplied by foods derived from plants. The best sources are:

(a) Whole grain cereals.

(b) Whole wheat breads.

(c) Fresh and dried fruits.

**NOTE**: Vegetables also help prevent constipation, but the fiber from whole grain cereals and whole wheat bread is more effective than vegetables.

3. **CALORIC REQUIREMENTS.** (FMST.07.01a) Recommended caloric intake (Naval Medical Command Instruction 10110.1) for troops operating in a cold weather environment is 4,500 calories per day.

a. Enemy Expenditure. Military operations in a cold weather environment usually involve strenuous activity. Walking on snow expends roughly twice the energy that walking on hard ground does. If troops are unfamiliar with personal over-snow mobility equipment, i.e. snowshoes and skis, awkward movement and thus wasted energy will result.

(1) The heavy clothing necessary for functioning in the cold increases energy expenditures by 5-15%. You will have the opportunity to experience this effect in the upcoming field evolution.

(2) Properly clothed, experienced individuals should have negligible heat loss from radiation, conduction, and convection. However, losses associated with respiration and evaporation can be significant.

(3) In intense cold, particularly with high winds, heat loss to the environment may be inevitable.

4. **NUTRITION IN A HIGH ALTITUDE ENVIRONMENT.** A diet high in carbohydrates and high in fat is the most beneficial, while a high protein diet is generally the least desirable.

a. The marked weight loss seen in individuals at high altitude is largely due to increased metabolic demands and anorexia secondary to hypoxia. Total caloric intake commonly does not meet energy demands.

b. Altered metabolic processes that have been reported at high altitude include delayed gastric emptying, intestinal malabsorption and decreased gastric enzyme secretion.

5. **BEST DIET FOR COLD WEATHER**. (FMST.07.01b)

a. High carbohydrate and high fat foods should be included rather than high protein foods.

b. Small meals (snacks) spaced at time intervals of two hours or less should be consumed rather than the usual three meals-a-day.

c. During a typical eight-hour day of cold exposure, three high-fat (unsaturated fat) meals providing 60% of the day's calories, coupled with two normally scheduled meals (breakfast and dinner) should provide good cold tolerance.

6. **BASIC CONSTITUENTS OF FOOD**. The three basic constituents of food are carbohydrates, fats and proteins.

a. Carbohydrates. Also known as the quick energy foods that provide energy to produce heat with the byproducts of carbon dioxide and water when oxidized by the body.

(1) Carbohydrates are present in food mainly as sugars and starches and are broken down during digestion into simple sugars that are converted into glucose.

(2) Carbohydrates are stored as glycogen in the liver and muscles and can be broken down quickly into glucose to provide a rapid source of energy. However, these stores are not large and are markedly depleted by fasting for as little as 24 hours.

(3) Eating a high carbohydrate diet for several days will double one's glycogen stores and can increase endurance by up to three times compared to the ordinary balanced diet.

(4) The main carbohydrate food sources are:

CARBOHYDRATE SOURCES

(a) Fruits.

(b) Vegetables.

(c) Cereals.

(d) Sugar.

b. Fats: In the body, fat serves as the main storage form of energy. One gram of fat has the equivalent of 9.3 Cal/gm as opposed to 4.1 Cal/gm for carbohydrates and proteins. They also produce energy, heat, $CO_2$ and $H_20$.

(1) In cases of starvation, body fat tends to be broken down into acidic compounds. If these compounds accumulate in the blood faster than they can be burned, they cause the body tissues and blood to become excessively acidic (acidosis).

(2) During brief physical activity, energy is derived equally from fat and carbohydrates, but as the duration of the activity lengthens, the percent of energy supplied by fat increases as carbohydrate energy is quickly depleted.

(a) Fat sources:

FAT SOURCES

1. Butter.
2. Lard.
3. Cooking oil.
4. Mayonnaise.
5. Ice cream.

(4) Fat is also found in lesser amounts in:

(a) Dairy products. (d) Nuts.

(b) Meat. (e) Vegetables.

(c) Eggs. (f) Cereals.

c. Proteins: Protein is a reparative food of complicated molecules composed of chains of amino acids.

(1) The protein structure is made up of 25 different amino acids. Eleven of these are important for growth while eight are essential for normal tissue maintenance.

(2) Common protein food sources are:

PROTEIN SOURCES

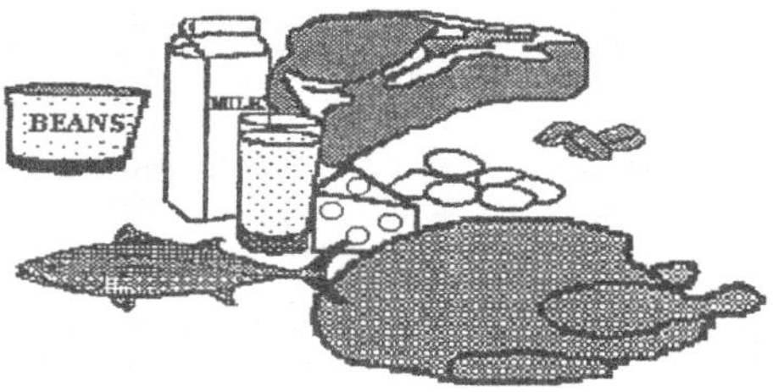


(a) Eggs.

(b) Dairy products.

(c) Meat.

(d) Poultry.

(e) Fish.

(f) Legumes (peas and beans, nuts and cereals with animal products, especially eggs, being better sources of complete protein).

(3) A pure protein diet may cause fatalities in 3-8 weeks from "rabbit starvation". This refers to those who have tried to live on relatively fat-free rabbit meat only. Fat must be a part of a normal diet.

7. **PERCENTAGE OF FATS, PROTEINS AND CARBOHYDRATES IN A SUMMER MOUNTAINOUS ENVIRONMENT**.

a. Requirements:

(1) 20% Protein.

(2) 35% Fats.

(3) 45% Carbohydrates.

b. When troops move to higher altitudes and/or are doing strenuous activity, carbohydrates should definitely be increased to at least 70 - 80% of energy intake. (Experienced mountaineers find carbohydrates more palatable at altitude).

8. **VITAMINS.** Vitamins are essential to the metabolic functioning of the body and are even more important in a cold weather environment because of an increase in metabolism and stress. Because the body cannot make vitamins, they are supplied in the food we consume. A daily multivitamin tablet supplement is recommended in a cold weather environment, but mega doses (more than 10 times the RDA) may be harmful.

a. Water-soluble vitamins are not stored in the body and need to be consumed daily. They are not stable to cooking and oxidation.

(1) B complex, Niacin, Pantothenic acid, Biotin, Choline, Folic Acid and C. (Note: Vitamin C facilitates absorption of iron from food.)

b. Fat-soluble vitamins can be stored in the body and are more stable to cooking and oxidation.

(1) A, D, E and K.

9. **FOODS THAT MAKE YOU FEEL WARM**.

a. Hot Beverages. Increases peripheral vasodilatation.

b. Adequate Dietary Iron. Mild iron deficiency increases chilling, thus decreasing cold tolerance.

c. Fat in the Diet. Fat slows gastric emptying and anecdotal evidence suggests less chilling and more restful sleep at night.

d. Carbohydrates. Keeps energy levels up. "Tired" is associated with "Cold".

e. Spices. Red pepper (Tabasco sauce) and monosodium glutamate increase peripheral vasodilatation, while capsicum and capsaicin, which are found in Cayenne, produces the same result when applied externally to the feet.

f. Water. Dehydration leads to lethargy, which tends to reduce food intake resulting in less heat production.

10. **INCREASING COLD TOLERANCE**.

a. Frequent Feedings. A high carbohydrate diet is superior to a high protein diet in increasing tolerance to cold.

(1) Diets deficient in both calories and protein reduce physical work capacity in the cold. The provision of both components is required to maintain work capacity.

11. **HIGH FAT DIETS AND SUSCEPTIBILITY TO FROSTBITE**.

a. High fat diets consumed at frequent (2 hour) intervals, maintain body temperatures better than high carbohydrate diets, but not if normal (4 hour) intervals are observed between meals.

b. Fat enriched diets lead to a propensity toward increased arterial thrombosis and increased platelet aggregation, which translates into increased potential for thrombosis.

c. Frostbite injuries are associated with disruption of the microcirculation, progressive micro vascular thrombosis, alteration of platelet function, and fibrin deposition in affected areas.

d. Therefore, diets high in (saturated) fat may predispose to frostbite.

**CAUTION!** A high-fat diet is important in a high-altitude, cold-weather environment. The benefits associated with the high-energy source called Fat far outweigh the risks of arteriosclerosis from the diet itself. A high-fat diet is appropriate in this specific environment during the time frame of cold, high-altitude exposure. In garrison, at sea level, a high fat diet should be avoided in order to maintain a healthy lifestyle.

12. **IRON DEFICIENCY AND THERMOREGULATION**.

a. Iron deficiency results in 9% less heat production.

b. Consuming only 1/3 of the RDA for iron results in a 29% greater heat loss during cold exposure. "The furnace is going, but the windows are open" is a phrase that helps describe this effect.

c. Iron absorption is increased by consuming iron rich foods:

(1) Lean meat.

(2) Poultry.

(3) Beets.

(4) Beans.

(5) Green leafy vegetables.

(6) Fish.

(7) Vitamin C.

d. Bottom line, IRON improves the body's response to cold.

13. **WATER REQUIREMENTS**.

a. On average, water makes up about 60% of the one's entire body weight. The average sedentary person excretes about 2.7 quarts of water each day. Of this total, 1.3 quarts are lost in the urine, 1.1 quarts through the skin and lungs and 10 ounces in the stool.

b. At high altitude, and during strenuous exercise, the amount of water lost through the skin and lungs due to the cold, dry air increases greatly.

c. Cold weather decreases the sense of thirst, which may lead to a state of chronic, mild dehydration. At temperatures below freezing and at elevations above the snow line, the lack of liquid water and the time and effort required to melt snow compound the problem.

d. Prevention of excessive water loss: (FMST.07.01c)

(1) Force fluids. Drink 6-8 quarts of water daily.

(2) Observe for dark urine and other symptoms of dehydration. (Headache, nausea/vomiting, dizziness, constipation, etc.)

(3) Limit caffeine to 1-2 cups per day.

(4) Do not rely on thirst as an indicator of hydration status.

**NOTE**: Eating snow or ice will cause extreme inflammation of the mucous membranes of the mouth after two days. Warm canteens of water placed in your sleeping bag will give you water to drink in the morning. They will initially act as hot water bottles. (Place them upright, as they tend to leak ...keep bag dry.)

e. Health Problems Associated with Excessive Water Loss:

(1) Dehydration.

(2) Constipation.

(3) Hypothermia.

(4) Frostbite.

(5) Heat Exhaustion.

f. The importance of hydration is essential while in a cold weather environment. If these guidelines are followed then complications arising from poor hydration should be prevented.

14. **<u>SPECIAL PURPOSE MILITARY RATIONS</u>**.

| RATION | APPLICATION |
|---|---|
| Ration, Cold Weather, RCW. | Cold weather operations where weight and freezing are a problem. |
| MRE Supplement Packet. | Cold weather calorie supplements for the MRE. |
| Ration, Light Weight, RLW. | Long range patrols, reconnaissance, and assault. |
| Nutrition Sustainment Module, NSM. | Calorie dense food, bar ration for mission lasting 3-5 days where soldier must carry his own food. |
| Family of Operational Rations, FOR. | Modular rations that allow an optimum mix of meals for tactical operations under all climatic conditions. |

15. **FIELD STRIPPING OF MRE'S.**

a. Although we talked briefly about MRE's, we did not really tell you how much to carry and how much to eat. Basically, if you were to carry enough MRE's, your pack weight could be excessively large. Here we offer you tips on how to conserve space and weight.

(1) Remove the entire contents from the plastic wrapper.

(2) Remove the cardboard, and all materials that you will not use.

(3) Be sure to place the toilet paper and other materials from the accessory pack either back in your pocket or into the brown packaging.

(4) Carefully repack the stripped meal back into the wrapper and seal with tape.

(5) Your meal will be half its original size and two-thirds its original weight.

**UNITED STATES MARINE CORPS**
Mountain Warfare Training Center
Bridgeport, California 93517-5001

FMST.07.11
10/23/01

## HYPOTHERMIA/REWARMING

**TERMINAL LEARNING OBJECTIVE**: Given a simulated casualty, treat a hypothermia casualty, in accordance with the references. (FMST.07.11)

**ENABLING LEARNING OBJECTIVES**:

(1) Without the aid of references, select from a given list the definition of hypothermia, in accordance with the references. (FMST.07.11a)

(2) Without the aid of references, select from a given list, the correct five mechanisms by which the body loses heat, in accordance with the references. (FMST.07.11b)

(3) Without the aid of references, select from a given list the reason why gentle handling of a hypothermia victim is important, in accordance with the references. (FMST.07.11c)

(4) Without the aid of references, select from a given list the four rewarming techniques used to rewarm a hypothermia victim, in accordance with the references. (FMST.07.11d)

(5) Without the aid of references, select from a given list the proper steps of the sleeping bag rewarming method of rewarming a hypothermia victim, in accordance with the references. (FMST.07.11e)

13

1. **HYPOTHERMIA**:

a. Definition: A fall in core body temperature to 95°F (35°C) or less. (FMST.07.11a)

b. Hypothermia is potentially lethal. The mortality rate is greater than 50% in severe cases and cases complicated by injuries or previous illness. The most common cause of death in hypothermia is ventricular fibrillation.

c. When speaking of hypothermia, core temperature values must be used. The term "exposure" has often been used to speak of hypothermia but this term is inexact and has numerous other connotations. It should not be used as a synonym for hypothermia.

d. Incidence.

(1) Classically, one thinks that hypothermia occurs only in extremely cold environments. However, it can occur in temperate climates and at Southern latitudes. Greater than 40% of cases of hypothermia occurred at ambient temperatures of 50°F or greater. Contrary to popular belief, the highest incidence of hypothermia is seen in urban areas.

(2) Hypothermia tends to be a male disease, with a male:female ratio of 2.8:1.0. It's also more frequent in the elderly.

(3) Hypothermia is often not recognized, as standard clinical thermometers don't register a low enough temperature. Currently, a low-reading rectal thermometer is available through the Federal Stock System (NSN 6515-00-139-4593). All medical personnel of cold weather units should carry this thermometer.

2. **PATHOPHYSIOLOGY**:

a. Heat Conservation/Loss.

(1) The body is a furnace that both generates and loses heat. An imbalance of these two processes can lead to hypothermia. About 75% of food energy consumed goes toward generating body heat while the other 25% goes toward maintaining basic metabolic function.

(2) Thermogenesis can be divided into two main categories:

(A) Shivering thermogenesis creates heat by muscle activity with subsequent breakdown of glycogen. Shivering usually begins at a core temperature of around 97°F (36°C) and finally ceases at core temperatures of 85°- 90°F (30°- 34°C).

(B) Non-shivering thermogenesis creates heat by hormones and their metabolic substances. Epinephrine, norepinephrine, growth hormone and thyroid hormone have all been implicated in playing a role in non-shivering thermogenesis.

(3) Changes in hormone secretion patterns are extremely important in acclimatization to cold weather climates.

(4) Once heat is generated it is conserved in the body by three main mechanisms: Peripheral vasoconstriction, abolition of sweating and piloerection. Peripheral vasoconstriction is by far the most important mechanism of heat conservation. The main internal temperature sensors in the body are located in the hypothalamus. When this area senses a drop in core temperature it initiates peripheral vasoconstriction at the surface of the skin to help preserve core body temperature.

(5) The body loses heat by five main mechanisms: (FMST.07.11b)

(A) Radiation is heat transferred to the surrounding environment via infrared radiation.

(B) Convection is the loss of heat via air/water current.

(C) Conduction is the direct contact-transfer of heat from a hot to a cold object.

(D) Evaporation is the conversion of water from a liquid phase to a gaseous phase (perspiration).

(E) Respiratory heat loss takes place when warm humidified air from the lungs is exhaled into the atmosphere.

(F) The mechanism accounting for the greatest percentage of heat loss is highly variable and depends on ambient conditions, i.e., a nude man standing in a cold room will lose approximately 65% of his heat by radiation. The same man standing in calm cold water will lose most of his heat by conduction and at a rate 25 times faster.

**NOTE: The liver and brain are the most thermogenic organs on a per weight basis. However, due to their greater mass, skeletal muscle and skin produce the largest percentage of body heat.

(6) Numerous factors predispose to the development of hypothermia. These include but are not limited to:

(A) Malnutrition.
(B) Lack of adaptation.
(C) Inactivity.
(D) Sedatives.
(E) Physical exhaustion.

(F) Burns.
(G) Extremes of age.
(G) Dehydration.
(H) Comorbid illness.

3. **SIGNS AND SYMPTOMS**:

a. Clinical manifestation of hypothermia reflect CNS, and cardiorespiratory involvement.

(1) 98°-95°F (37°-35.6°C): Mild shivering, cold sensation and impaired fine motor coordination.

(2) 95°- 92°F (35°-32.8°C): Violent shivering, difficulty speaking, sluggish thinking, amnesia, large muscle incoordination.

(3) 92°-86°F (32.2°-30°C): Shivering is replaced by muscular rigidity. Exposed skin is blue or puffy. Movements are jerky. Dulled sensorium, but victim may be able to maintain posture and the appearance of being in contact with surroundings. Possible atrial arrhythmias.

(4) 85°-81°F (29.4°-27.2°C): Coma, lack of reflexes, possible ventricular arrhythmias. Cessation of shivering below 86oF.

(5) Below 78°F (25.6°C): Failure of cardiac and respiratory centers, pulmonary edema, ventricular fibrillation, asystole. Death.

b. Some patients with hypothermia will exhibit a phenomenon that is called, "paradoxical undressing."

(1) In these instances the hypothermia victim is found nude with their clothes beside them. If the victims are women they are often misdiagnosed as sexual assault cases. The pathophysiology is that as the body's core temperature continues to drop, hypothalamic control of peripheral vasoconstriction is lost, flushing the periphery with relatively warm blood causing the individual to feel flushed and warm. Subsequently the individual undresses to cool down.

c. Again it must be stated that the only way to correctly diagnose hypothermia is with a core temperature reading.

4. **FIELD TREATMENT**:

a. ABC's, C-Spine precautions, remove wet clothing, replace with dry clothes and insulate with a vapor barrier system. (Wrapping the victim in two plastic garbage bags or bubble wrap is ideal. Further insulation may be necessary). Handle the victim carefully to prevent ventricular fibrillation caused by unnecessary jarring and rough handling. (FMST.07.11c) CASEVAC.

b. Victims with altered consciousness should be NPO. Under no circumstances should ETOH be given. This is a peripheral vasodilator and will only enhance heat loss.

c. If CASEVAC is not feasible for a long period of time, or if the victim is only mildly hypothermic, it may be necessary to rewarm in the field by using a proper sleeping bag rewarming technique. Additionally, insulated hot packs placed at the high heat loss areas will help.

d. The hypothermic victim often has no discernible vital signs. This brings up the question as to whether or not to administer CPR.

(1) Expert opinion is divided on this matter. One school states that since no vital signs are present, CPR should be done. The other school states that to do CPR will probably do more harm than good because the victim often has adequate circulation even though it may not be discernible and CPR may send them into ventricular fibrillation.

(2) Be aware that severe hypothermic victims in ventricular fibrillation and asystole have been successfully resuscitated even after periods as long as 4 hours.

5. **REWARMING**: The re-warming of a hypothermia casualty in a BAS setting is less than optimal since the AMAL for a BAS does not include monitoring devices, laboratory facilities, etc., but rewarming of a casualty can be successfully conducted even in these fairly primitive conditions. The following are three basic techniques of re-warming:

- Passive external re-warming: a sleeping bag or blanket that is not pre-warmed.
- Active external re-warming: a pre-warmed sleeping bag, bear hugger or warm water bath.
- Active internal or core re-warming: IV fluids, warmed/humidified O2, lavage, extracorporal re-warming.

a. Four ideal re-warming strategies for the BAS: (FMST.07.11d)

(1) Sleeping bag rewarming.

(2) Water bath/counter current rewarming.

(3) Warm/humidified airway rewarming.

(4) Warmed I.V. solutions. NS is preferable as a cold liver does not like to metabolize lactate.

6. **REWARMING TECHNIQUES (BAS)**:

a. Sleeping Bag Rewarming. (FMST.07.11e) This is the simplest, but least effective method.

(1) Make the diagnosis. Take a rectal temperature to determine if the casualty is clinically hypothermic (95°F or less). Use this temperature as a base line to determine stabilization.

(2) Warm sleeping bag (pre-warm sleeping bag with two volunteers). Placing a hypothermic casualty into a cold sleeping bag will cause further heat loss by conduction.

(3) Strip casualty. Remove all wet clothing avoiding unnecessary handling. Muscular movement will pump cold blood to the core.

(a) Place the casualty between the two volunteers, if space permits. Their body heat is transferred from the two volunteers to the cold casualty.

(4) Monitor core temperature frequently throughout transport to ensure that the casualty is stabilized and not going into deeper hypothermia.

(5) Adequate insulation is required to prevent further heat loss to the environment.

(6) Augment heat by placing insulated heating pads in the high heat loss areas: head, axilla, groin, popliteal region and antecubital fossa. (Be careful not to cause burns)

**NOTE: It should be noted that an evacuation bag can be used, this will provide more room for the casualty and volunteers.

b. Water Bath Rewarming. This method has a long history of success in rewarming even severely hypothermic casualties. The casualty is placed in a portable/field expedient bathtub or a life raft. The extremities and head are kept out of the water to avoid afterdrop. The water bath temperature should be 104°-108°F. An anesthesia temperature probe should be used to constantly monitor the core temperature. Since the casualty will cool the bath water around him, it will be necessary to continue to add warm water to maintain the proper water bath temperature.

**NOTE: Individual in heat distress can be rapidly cooled using this method with "cool" water instead.

c. Warm Airway Rewarming A warmed and humidified air/oxygen mixture is used to provide a warming media within the lungs. These devices consist of a method of generating warm humidified air and are usually portable. The warm humidified air moderately increases the amount of heat that can be delivered to core. There should be a thermometer in the airway tubing to monitor the temperature. To prevent injury to the bronchi, the temperature should not exceed 115°F (108° to 115°F is ideal). Warm airway rewarming can also be used in conjunction with other rewarming methods. This method, by itself, will probably not deliver sufficient heat to rewarm a severely hypothermic casualty by itself.

Caveat: When possible, active external re-warming should always be combined with active core re-warming techniques so as to minimize the effects of afterdrop. This occurs as a result of peripheral vasodilation causing warm core blood to move to the surface while relatively cold surface blood moves to the core.

**NOTE: Hypothermic submersion incident casualties cannot tolerate humidified air for any length of time.

d. Warmed I.V. Solutions. Hypothermia is a common response to I.V. therapy. Solutions that have been prewarmed have been shown to prevent this complication. There are various methods for warming I.V. solutions:

(1) Crystalloid solutions can be warmed by warm water bath and microwaved with no adverse changes to its integrity.

(2) Fresh Frozen Plasma (FFP; -20°C): Microwaving and water bath rewarming have showed an increase in plasma hemoglobin from residual RBC's. By microwaving, you will have an 85.7% increase, while in a water bath there was only a 53% increase. However, by microwaving for 30 seconds then manipulating the bag for 10 seconds, with the process being repeated five times you have only 2.6% cell destruction.

(3) Packed Red Blood Cells (PRBC 4°C): Using conventional rewarming methods, studies have shown that the outer five millimeters of the bag can have isolated areas of hemolysis. Until a device is made that can uniformly shake up PRBC during warming, we do not recommend microwaving of blood. Rewarming of blood by dilution with warm, calcium-free crystalloid solution is useful.

WIND CHILL CHART

| WIND | AMBIENT TEMPERATURE (FAHRENHEIT/CENTIGRADE) | | | | | | | | | | | | | |
|---|---|---|---|---|---|---|---|---|---|---|---|---|---|---|
| CALM F | 35 | 30 | 25 | 20 | 15 | 10 | 5 | 0 | -5 | -10 | -15 | -20 | -25 | -30 |
| CALM C | 27 | -1 | -4 | -7 | -9 | -12 | -15 | -18 | -21 | -23 | -26 | -29 | -32 | -34 |
| EQUIVALENT WIND CHILL FACTOR (FAHRENHEIT/CENTIGRADE) | | | | | | | | | | | | | | |
| 5 MPH F | 33 | 27 | 21 | 16 | 12 | 7 | 1 | -6 | -11 | -15 | -20 | -26 | -31 | -35 |
| 8 KPH C | 1 | -3 | -6 | -9 | -11 | -14 | -17 | -21 | -24 | -26 | -29 | -32 | -35 | -37 |
| 10 MPH F | 21 | 16 | 9 | 2 | -2 | -9 | -15 | -22 | -27 | -31 | -38 | -45 | -52 | -58 |
| 16 KPH C | -6 | -9 | -13 | -17 | -19 | -23 | -26 | -30 | -33 | -35 | -39 | -43 | -47 | -50 |

| | | | | | | | | | | | | | | |
|---|---|---|---|---|---|---|---|---|---|---|---|---|---|---|
| 15 MPH F | 16 | 11 | 1 | -6 | -11 | -18 | -25 | -33 | -40 | -45 | -51 | -60 | -65 | -70 |
| 23 KPH C | -9 | -12 | -17 | -21 | -24 | -28 | -32 | -36 | -40 | -43 | -46 | -51 | -54 | -57 |
| 20 MPH F | 12 | 3 | -4 | -9 | -17 | -24 | -32 | -40 | -46 | -52 | -60 | -68 | -76 | -81 |
| 32 KPH C | -11 | -16 | -20 | -23 | -27 | -31 | -36 | -40 | -43 | -47 | -51 | -56 | -60 | -63 |
| 25 MPH F | 7 | 0 | -7 | -15 | -22 | -29 | -37 | -45 | -52 | -58 | -67 | -75 | -83 | -89 |
| 40 KPH C | -14 | -18 | -22 | -26 | -30 | -34 | -38 | -43 | -47 | -50 | -55 | -59 | -64 | -67 |
| 30 MPH F | 5 | -2 | -11 | -18 | -26 | -33 | -41 | -49 | -56 | -63 | -70 | -78 | -87 | -94 |
| 48 KPH C | -15 | -19 | -24 | -28 | -32 | -36 | -41 | -45 | -51 | -53 | -57 | -61 | -66 | -70 |
| 35 MPH F | 3 | -4 | -13 | -20 | -27 | -35 | -43 | -52 | -60 | -67 | -72 | -83 | -90 | -98 |
| 56 KPH C | -16 | -20 | -25 | -31 | -33 | -37 | -42 | -47 | -51 | -55 | -58 | -64 | -68 | -72 |
| 40 MPH F | 1 | -4 | -15 | -22 | -29 | -36 | -45 | -54 | -62 | -69 | -76 | -87 | -94 | -101 |
| 64 KPH C | -17 | -20 | -26 | -30 | -34 | -38 | -43 | -48 | -52 | -56 | -60 | -66 | -70 | -74 |

## 7 REWARMING TECHNIQUES (HOSPITAL):

a. The best re-warming techniques to be used at higher echelon care facilities still tend to employ rather simple therapeutic modalities. The three safest and most effective are listed below:

- Warmed IVF.
- Warmed/humidified oxygen.
- Bear Hugger.

b. Many texts advocate the use of extracorporal blood re-warming. However, this technique has some disadvantages. It requires a high level of technical and medical support, which are often not available. Since patients undergoing bypass must be anti-coagulated, patients with multiple traumatic injuries may not be good candidates. Lastly, research indicates that anticoagulation with heparin may actually worsen the severity of frostbite, which may occur in patients with hypothermia.

c. Lavage: peritoneal, thoracic and mediastinal lavage have all been suggested for use in severe hypothermia. However, the use of these techniques in inexperienced hands tends to carry a much higher infection and complication rate. Gastric, colonic and bladder lavage are indeed safer but are of limited efficacy as they do not offer access to areas with significant surface area.

**UNITED STATES MARINE CORPS**
Mountain Warfare Training Center
Bridgeport, California 93517-5001

FMST.07.10

10/25/01

**STUDENT HANDOUT**

# FREEZING AND NEAR FREEZING TISSUE INJURIES

**TERMINAL LEARNING OBJECTIVE**: Given a simulated casualty, treat common cold injuries, in accordance with the references. (FMST.07.10)

**ENABLING LEARNING OBJECTIVES**.

1) Without the aid of references, from a given list select the correct definition of frostbite, in accordance with the references. (FMST.07.10a)

2) Without the aid of references, from a given list select the correct two mechanisms of injury for frostbite, in accordance with the references. (FMST.07.10b)

3) Without the aid of references, from a given list select the correct signs and symptoms of frostbite, in accordance with the references. (FMST.07.10c)

4) Without the aid of references, from a given list select the correct favorable post thaw signs of a frostbite injury, in accordance with the references. (FMST.07.10d)

5) Without the aid of references, from a given list select the correct unfavorable post thaw signs of a frostbite injury, in accordance with the references. (FMST.07.10e)

6) Without the aid of references, from a given list select the correct field treatment for frostbite, in accordance with the references. (FMST.07.10f)

**14**

7) Without the aid of references, from a given list select the correct primary field treatment consideration when treating frostbite in the field, in accordance with the references. (FMST.07.10g)

8) Without the aid of references, from a given list select the correct four methods to reduce the chance of becoming a frostbite casualty, in accordance with the references. (FMST.07.10h)

9) Without the aid of references, from a given list select the correct definition of immersion foot, in accordance with the references. (FMST.07.10i)

10) Without the aid of references, from a given list select the correct mechanism of injury for immersion foot, in accordance with the references. (FMST.07.10j)

11) Without the aid of references, given a list of categories and characteristics match the category of immersion foot to its characteristics, in accordance with the references. (FMST.07.10k)

12) Without the aid of references, from a given list select the correct field treatment of immersion foot, in accordance with the references. (FMST.07.10l)

**OUTLINE**.

1. **FROSTBITE**

a. Definition: Frostbite is the actual freezing of tissue. (FMST.07.10a)

(1) Frostbite usually occurs as a result of exposure to subfreezing temperatures for a long period of time. However, brief exposure to high wind-chill can also cause frostbite. Frostbite usually occurs after more than twelve hours of exposure to subfreezing temperatures. Skin temperature of the extremities is dependent on heat production and is more a function of underlying muscle mass than of circulation. Thus, the areas of the body most prone to frostbite (hands, feet, ears, nose, chin and molar regions) have small amounts of underlying muscle tissue.

(2) Incidence. Frostbite, unlike hypothermia tends to spare the extremes of age. Thus, it is most prevalent in the middle years of life and like hypothermia, the greatest incidence occurs in urban areas. Other factors predisposing a person to frostbite include: poor nutritional status, dehydration, prior frostbite injury, individuals from the southeastern part of the United States, diabetics, individuals with atherosclerotic disease, and use of tobacco products.

(3) Although the actual numbers are not known, persons born in the south are 3.7 times more susceptible to frostbite. Persons with a history of prior cold injury are 1.7 times

more prone to have a second injury. This includes parts of the body other than the originally frostbitten region.

b. Mechanisms of Tissue Injury. Tissue injury resulting from frostbite is produced in two ways: (FMST.07.10b)

(1) *Actual freezing of the tissues.* Ice crystal formation in the extra-cellular space leads to an osmotic gradient across the cell membrane drawing water out of the cell. This process, known as crenation, causes a marked change in intra-cellular electrolyte and enzyme concentrations, which can lead to cell death or recovery depending on the duration of exposure to freezing conditions. Intra-cellular ice crystal formation is much more destructive, however, this requires much faster freezing than typically occurs with frostbite.

(2) *Obstruction of blood supply to the tissues.* Cell injury and death caused by ice crytal formation leads to the release of inflammatory mediators, which damage vascular endothelial cells. The net result is increased vascular permiability. Fluid leaks out of the vascular space and into the intercellular space causing tissue edema and increased blood viscosity. Elevated viscosity combined with platelet aggregation ,to the walls of damaged blood vessels, lead to vascular obstruction with resultant tissue damage.

c. Signs/Symptoms. (FMST.07.10c) Initially there is discomfort or pain in the affected tissues which can progress to loss of sensation. The skin appears white and bloodless. There is decreased or no capillary refill. The skin may feel waxy and will not glide freely over a bony prominence. With minor injury blisters form, usually in the first 24 hours and edema may also be present. Deep injury is not associated with substantial formation of blisters. There is also little edema seen in deep injury.

(1) Favorable prognostic signs of frostbitten tissue. These signs indicate probably little or no tissue loss, post thaw. (FMST.07.10d)

(A) Warmth of tissue.

(B) Normal tissue color.

(C) Preservation of sensation.

(D) Blister formation. Blisters extending all the way, distally in the affected part, are a particularly good prognostic sign, however, blisters limited to the line of demarcation between normal and affected tissues are a less favorable sign.

(E) Edema persisting more than 24 hours.

(F) Rapid capillary refill.

(2) Unfavorable prognostic signs. These indicate a poor prognosis with probable tissue loss, post thaw: (FMST.07.10e)

(A) Complete absence of edema.

(B) Cyanotic tissue.

(C) Continued loss of sensation.

(D) Affected areas continue to stay cold.

(3) Since it is extremely difficult to predict ultimate tissue loss from initial examination, any classification must be retrospective. Classically, frostbite is divided into four degrees of severity:

(A) First degree. Erythema, edema, transient tingling and/or burning.

(B) Second degree. Blisters, edema, anesthesia and/or paresthesias.

(C) Third degree. Involves the entire thickness of skin and extends into subcutaneous tissue.

(D) Fourth degree. Involves the entire thickness of the part, including bone.

(E) A more useful classification is:

1. Superficial frostbite - no tissue loss.

2. Deep frostbite - some tissue loss.

(F) Frostnip is a completely reversible injury which is technically not considered frostbite. Frostnip classically presents as a white patch and numbness and is readily warmed in the field by placing the affected extremity in one's axilla or groin. No resultant skin changes occur.

d. Field Management (FMST.07.10f)

(1) The correct treatment for frostbite is as follows:

(A) Once an extremity has suffered frostbite, the primary consideration of treatment is not to rewarm the region if there is a chance of refreezing. (FMST.07.10g) This occurs because the amount of tissue damage is greatly increased if the part is thawed then allowed to re-freeze. Generally, the sooner a frostbitten region is thawed the lower the probability and extent of damage. Rapid rewarming is the preferred method of thawing a frostbitten region. This consists of immersion of an affected part in warm water of 104°-108°F until the tissues are soft and pliable, which usually takes about 40-60 minutes.

(B) If slow rewarming methods (such as allowing a part to sit at room temperature) is used, some refreezing of the melted fluid occurs and the new ice crystals are actually larger than those formed during the original freezing causing further tissue destruction.

(C) Out dated methods such as rubbing with snow or rewarming next to an open fire, etc., only enhance tissue damage.

(D) Once thawing has taken place no benefit of continued rapid rewarming can be demonstrated.

(E) Cold injuries which are old and have thawed spontaneously should not be rapidly re-warmed again.

(F) Nicotine use should be prohibited since it is a vasoconstrictor and may enhance tissue loss.

(G) There is no need for ointments, salves, etc.

(H) Blisters should be allowed to remain intact.

(I) Medevac all cases of frostbite, as soon as the tactical situation permits.

(J) One final note; in military tactical situations such as ours, the victim often must walk down below the snow line to reach a BAS or base camp where rapid rewarming can be carried out. In this situation let the victim walk if he has frostbitten feet. Once his feet have thawed, he will be unable to walk and will have to be carried out.

e. <u>Prevention</u> (FMST.07.10h)

(1) Some general principles for the prevention of frostbite are listed below. (Adapted from Washburn's classic article in the NEJM).

(a) Dress to keep comfortably cool. The head radiates 30-50% or more of all body heat and so goes the old adage, "If your feet are cold put on your hat".

(b) Nutrition. Eat the right type and quantity of foods.

(c) Avoid tight fitting, constrictive clothing. This decreases blood flow and increases the chance of frostbite. Rings, jewelry, etc., conduct heat away from the skin and increase the chance of frostbite.

(d) Wear mittens instead of gloves. They are much warmer. For work involving a high level of dexterity use contact gloves.

(e) Do not touch metals with bare hands at extremely low temperatures. Your flesh will stick to the surface and this will result in subsequent frostbite and tissue loss. If you find yourself in such a predicament pour a warm fluid over the metal to thaw your flesh off the metal.

(f) Don't use nicotine.

2. **IMMERSION FOOT**.

a. Definition. (FMST.07.10i) Immersion foot (AKA trenchfoot) is a non-freezing injury of the extremities in which the tissues are damaged.

b. Usually a disease of the feet (but it can occur in the hands), it occurs when the extremity is exposed to a cold-wet environment (external temperatures above freezing and less than $50^{O}$F) for long periods of time (usually greater than 12 hours). Prolonged standing enhances the pathologic potential for injury due to wet and cold conditions.

c. Mechanism of Injury. (FMST.07.10j) Even though this injury is called immersion foot, actual immersion of the feet in water is not necessary. This injury can also happen to the hands.

(1) The circulation of blood to the affected extremity is reduced because the victim's extremity is cold, wet and the peripheral blood vessels have constricted to reduce overall heat loss.

(2) Because prolonged cooling damages all tissue, nerve injuries are also involved. This nerve damage is responsible for the severe pain and paresthesias. Damaged skin and subcutaneous ttissues result in erythema, edema and possible blister formation.

d. Signs/Symptoms Early signs include numbness and paresthesias followed by complete numbness and sometimes leg/arm cramps. The skin at first appears erythematous but becomes progressively pale and mottled then grayish-blue as the condition progresses. Usually there is a large amount of edema and blisters. In severe cases the arterial pulses are absent. Immersion foot is classified by severity: (FMST.07.10k)

| CATEGORY | CHARACTERISTICS |
|---|---|
| Minimal | Reddening of skin; slight sensory changes |
| Mild | Edema; sensory changes (reversible) |
| Moderate | Edema, redness, blisters, intra-cutaneous hemorrhage; irreversible nerve damage |
| Severe | Severe edema, massive intra-cutaneous hemorrhage; necrosis, gangrene |

e. Field Management (FMST.07.101) Treatment consists of:

(1) Pat drying of extremity.

(2) Gentle rewarming.

(3) Elevation of affected extremity.

(4) Bed rest.

(5) Treatment after this is supportive. For more severe injuries, debridement or resection of a portion of the foot or even amputation may be necessary. However, this is best accomplished at a higher echelon of care.

3. **CHILBLAINS (PERNIO).**

a. Definition. Chilblains are a non-freezing injury that leads to red inflamed, pruritic skin.

b. Mechanism of Injury. Pernio is caused by intermittent exposure to above freezing temperatures and high humidity. (More extreme cold is usually associated with lower humidity). Classically, chilblains occurs in young individuals and most commonly on the anterior surface of the legs and the dorsum of the hands. The cause of pernio is not clearly known, but it is thought to reflect an abnormal vascular response to the environment.

c. Signs/Symptoms. Skin is red, tender, pruritic, edematous and blanches with pressure. Itching may be severe and hyperhydrosis can occur.

d. Management. The field treatment is massage and gentle heat. This treatment is usually self-limiting and will not recur without re-exposure. Topical steroids are effective for itching and burning.

4. **CORNEAL FROSTBITE**

a. Definition. Actual freezing of the cornea.

b. Mechanism of injury. This occurs when large amounts of cold air pass over and cool the cornea, resulting in corneal clouding and loss of vision. This can occur two ways, either skijoring or while driving a snow mobile without wearing proper eye protection (ski goggles).

c. Signs/Symptoms. On examination the cornea will be clouded over.

d. Treatment and Prevention. The only treatment available is corneal transplantation. Prevention is the key.

**UNITED STATES MARINE CORPS**
Mountain Warfare Training Center
Bridgeport, California 93517-5001

LP
10/23/01

**STUDENT HANDOUT**

**EXTREME COLD WEATHER TENT**

**PURPOSE**. The purpose of this class is to familiarize you with the Extreme Cold Weather Tent, its components, uses, and maintenance.

**OUTLINE.**

1. **CHARACTERISTICS AND NOMENCLATURE**

   a. Characteristics. The extreme cold weather tent was developed to replace the Norwegian tent sheets. It is lightweight and portable, weighing only 14 lbs. The tent is a self-standing, dome-shaped, four season design capable of holding four Marines within its 84 square feet of floor space.

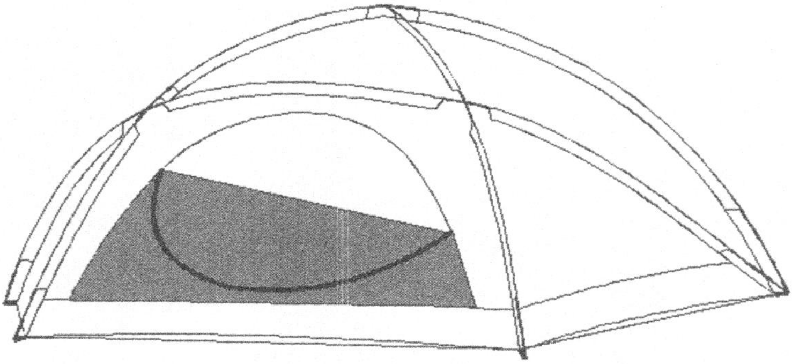

**EXTERNAL VIEW**

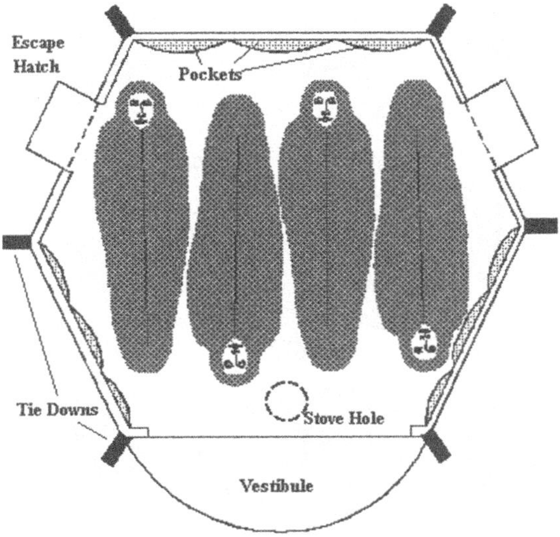


INTERNAL NOMENCLATURE

b. Nomenclature.

(1) Tent body. Made of 3 ounce per yard urethane coated taffeta nylon. Inside the top of the tent is a mesh drying rack, and around the bottom are several mesh pockets for commonly used items. The entrance has a no-see-um mesh panel designed to keep bugs out. Later versions will have two openings spaced around the tent body that facilitate joining tents together for CP's or BAS's. Another modification on later models is a stove hole by the front door that closes with a drawstring. This allows cooking to be done inside the tent during inclement weather without fear of food spills soiling the tent or fire igniting it in case the stove falls over.

(2) Flysheet. The tent comes with three different flysheets; a woodland camo cover for forested areas, a desert camo cover for deserts, and a white cover for snow-covered terrain. These sheets are also made of nylon with a heavier urethane coating. Later models of this tent will only come with two flysheets, the desert one having been deleted.

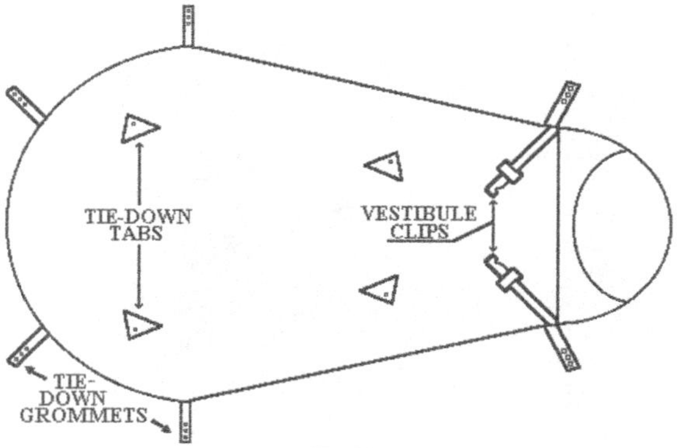


**FLYSHEET**

(3) Poles. The pole configuration used with this tent allows maximum use of floor space. The poles are comprised of nine sections of 7075 aluminum with an overall length of 18 ft 2 1/2 in. The poles are held together by shock cords, which aid in connecting them when pitching the tent. Later models of the tent will have slightly shorter poles that are easier to put into the sleeves and reduce bending the poles by forcing them.

(4) Accessory Kit. Each tent comes with an accessory kit containing; 2 pole repair sleeves, 24 aluminum stakes, 12 nylon tie down cords, 12 line tighteners, a black foam spacer, and woodland colored repair tape 3" x 36".

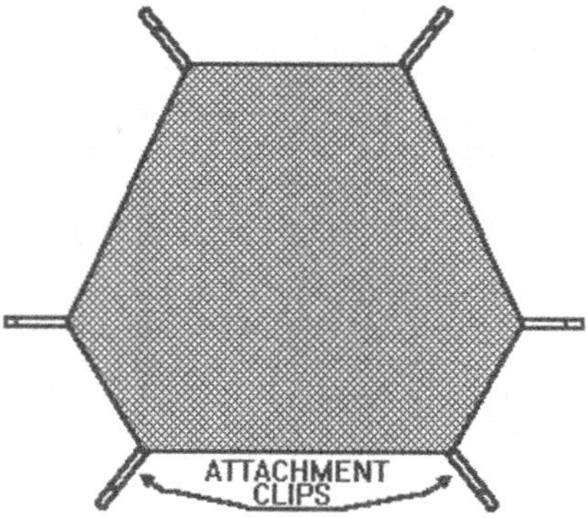


**MESH DRYING RACK**

2. **PITCHING THE TENT**

a. Clear an area. Ensure that there is sufficient room for the tent (approximately 12 feet) by spreading it out on the ground and pulling the floor section tight. Another method is to have one-man stand in the center of where the tent is to be pitched and hold a ski pole for a second Marine. The second man will walk in a circle around the first man using the ski pole to measure the radius, thus ensuring that the tent will have enough room. Once the circle is marked, should be dug down (4 to 6 feet). Depending on the tactical situation, if time is limited, you can pack down the snow to achieve some cover and concealment initially and then improve the position later by building up a surrounding snow wall.

b. Insert poles into sleeves. Six of the poles go into the sleeves on the tent, with three being kept aside for the fly. The poles that form the triangle at the top of the tent should go in first, followed by the poles around the side. There are several grommets in each strap to adjust the tension of the tent. If the tent is too loose, snow and rain can accumulate.

***NOTE***: In pitching and striking, it is advisable to push rather than pull the poles so that the sections will remain engaged.)

c. Foam Spacer. Attach spacer to snap located on the rear of the tent. The foam spacer is used to prevent the tent sheet from coming into contact with the tent, keeping water away from the tent.

d. Flysheet. The fly is hooked onto the back of the tent and brought over the top, ensuring that the entrances on both are aligned. Insert the remaining poles into the sleeves and adjust for tension. Pull the front of the fly out away from the tent to attain maximum tension. Inside the fly are two straps that attach to the triangular buckles on each side of the entrance, use these to adjust tension and prevent the fly from blowing away. The fly is a very important part of the tent and tent performance is degraded without it, so it is imperative that the fly not be allowed to blow away.

e. Securing. Use the tent stakes and guidelines provided to secure the tent. These tents, as with all tent age, are vulnerable to wind damage; therefore, it may be necessary to secure the corners prior to inserting the poles during pitching in high wind conditions. If you are pitching the tent in deep snow, it may be preferable to use 'deaden' to hold the tent down. Use all tie down points available depending on wind and tactical conditions. (Tieing the tent down in ten different places may not be advisable if enemy attack is imminent.)

f. Tactical considerations. The tent is designed with a light retention material, but it is not lightproof. Also, it may be possible to build a snow wall that not only shields light emissions, but also camouflages and protects the tent from snow as well. In deep snow it is best to dig down into the snow pack and keep a low silhouette. Camoflauge with over whites or netting.

3. **STRIKING AND PACKING**. In order to strike the tent, perform the pitching instructions in reverse order. It is recommended that the tent fly be folded length ways into thirds. Before

placing it in stuff sacks, roll tightly around the folded tent pole sections, squeezing trapped air out in the process.

4. **MAINTENANCE**. The ECW tent requires very little maintenance.

   a. Cleaning. After each use, shake out loose debris. Sponge clean all dust and track marks.

      (1) If the fabric requires deeper cleaning, hand wash the tent in mild soap and warm water.

      (2) Air-dry the tent out of direct sunlight. Make sure that the fabric is completely dry.

      (3) Never store tent damp; it will cause mildew and damage to the tent fabric.

   b. Tent Pole Care. From time to time, apply a thin layer of silicon lubricant to all parts of the poles. This is excellent protection against corrosion, prevents the poles from freezing together when they are very cold, and will make the joints work more smoothly in any weather.

   c. Seam Sealing. To ensure waterproofness of both the tent floor and flysheet, the seams must be thoroughly sealed.

   d. Zippers. Lubricate the zippers with a silicone spray to keep them running smoothly and to prevent freezing.

5. **SAFETY CONSIDERATIONS**.

   a. Ventilation. With four people in the tent it can become a very foul atmosphere inside very quickly. On the original version of this tent the only way to ventilate it were to open the door slightly, later versions having positive ventilation in the sidewalls. This also prevents condensation from forming inside the tent and humid air preventing the proper functioning of the drying rack.

   b. Stoves. The place for cooking or melting snow for water is the vestibule. Great care should be taken when lighting a stove inside the tent because a flare up could be disastrous. If weather conditions prohibit cooking outside of the tent, then do so inside the vestibule. Cooking inside the tent itself can lead to fires or nasty spills on gear and other Marines resulting in burns. A small stove will heat the inside of the tent very quickly, but it will also consume all of the air in a sealed tent, resulting in asphyxiation. There have been numerous deaths caused by this. Also, you should never sleep in the tent while the stove is lit, this can have predictable results. If at all possible, do all cooking during the day as a light discipline technique.

6. **CARRYING THE ECW TENT**. The North face tent is designed to be used by a fire team and will be carried by the fire team to which it is assigned. There are a couple ways this will be accomplished. The first way is that the tent can be spread loaded between all the members of the fire team. The second way that the North face tent can be moved is by use of the

Armidilla sled. When using the sled the North face tent will be placed between the normal fire team stores and the pioneer gear, so that it is readily accessible when it needs to be used.

**UNITED STATES MARINE CORPS**
Mountain Warfare Training Center
Bridgeport, California 93517-5001

LP
10/23/01

**STUDENT HANDOUT**

**PERSONAL/TEAM STOVES**

**PURPOSE**. The purpose of this period of instruction is to familiarize the student on the operation and maintenance of the Optimus, MSR and Peak 1 stoves. This lesson relates to the ECW tent.

**OUTLINE.**

1. **NOMENCLATURE OF 8R OPTIMUS STOVE**. To be certain on how to operate any one item, you must first be able to identify the different parts and understand each part's task.

   a. Burner Assembly. The burner is the actual heart of the stove. Its main function is to regulate the flow of fuel from the fuel tank in through the burner and to direct flame upward.

   b. Cooking Rack. Supports item that is being cooked, i.e., canteen cup, twin pot cook set.

   c. Protecting Shield. Protects fuel tank from overheating, from the flame that is put out by the stove.

   d. Regulating Key. Controls fuel flow to the burner.

   e. Spirit Cup. Holds fuel that will be used if priming of stove is necessary.

   f. Pump Assembly. Builds pressure within the fuel tank forcing fuel through the wick to the burner assembly.

   g. Fuel Tank.

**16**

NOTE: Complete rotation of regulating key counterclockwise will move the cleaning needle through the nipple hole removing most carbon build-up from within the nipple hole.

2. **SERVICEABILITY CHECK FOR 8R OPTIMUS STOVE**. Check for obvious damage such as broken and bent parts. Ensure stove will work safely.

3. **LIGHTING THE 8R OPTIMUS STOVE**.

   a. Ensure that the fuel tank has sufficient fuel in it. No more than 3/4 full to allow for pressurization.

   b. Replace tank cap.

   c. Place pump assembly onto the tank cap.

   d. Gently pump the plunger until you feel sufficient pressure in the fuel tank.

   e. Strike a match.

   f. Turn regulating key counterclockwise just enough to let a little fuel escape.

   g. Take the lit match and place it near the burner head.

   h. Stove should light; use regulating key to adjust how high you want the flame.

   i. Priming. Due to coldness, priming may be necessary.

      (1) Place a little fuel in the spirit cup.

      (2) Light fuel.

      (3) Regulating key. Before lit fuel goes out, turn regulating key cc to let fuel escape.

      (4) Burner assembly. This will have warmed-up the burner assembly and the stove should now light.

4. **NOMENCLATURE OF 111B OPTIMUS STOVE**.

   a. Burner Assembly. The burner is the actual heart of the stove. Its main function is to regulate the flow of fuel from the fuel tank in through the burner itself and to direct flame upwards.

   b. Outer Spindle Wheel. Carries out the same task as the regulating key on the 8R Optimus stove.

   c. Spirit Cup. Same as 8R Optimus stove.

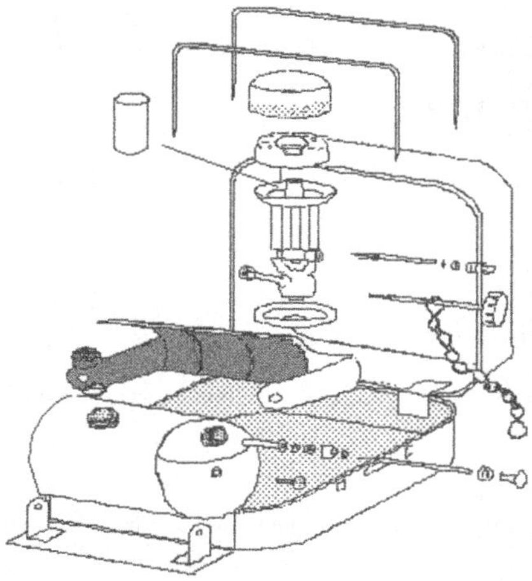

**OPTIMUS 111B**

d. Inner Cap. Its function is to become hot enough to vaporize the fuel as it leaves the burner assembly.

e. Outer Cap. Contains the flame and distributes it evenly in an upward direction.

f. Pump Rod With Piston Assembly. Creates the pressure within the fuel tank that enables fuel to be passed from the fuel tank to the burner assembly.

5. **SERVICEABILITY CHECK FOR 111B OPTIMUS STOVE**. Same as 8R Optimus stove.

6. **LIGHTING THE 111B OPTIMUS STOVE**.

a. Ensure that the fuel tank has sufficient fuel in it. No more than 3/4 full to allow for pressurization.

b. Replace tank cap finger tight.

c. Place pump assembly onto the tank cap.

d. Gently pump the plunger until you feel sufficient pressure in the fuel tank.

e. Strike a match.

f. Turn the outer spindle wheel counterclockwise just enough to let a little fuel escape.

g. Take the lit match and place it near the outer cap.

h. Stove should light; use spindle wheel to adjust the flow of fuel controlling the flame.

i. Priming. Due to coldness, priming may be necessary.

(1) Place a little fuel in the spirit cup.

(2) Light fuel.

(3) Before lit fuel goes out, take the outer spindle wheel and turn it counterclockwise to let fuel escape.

(4) This should light the stove.

***NOTE***: Again, priming usually will be needed under severe cold conditions or if the stove is frozen.

7. **NOMENCLATURE OF THE MSR WHISPERLITE STOVE**. In order to operate this stove efficiently, identification of the different parts and their functions is what you'll need to know.

a. Burner Head. This is the heart of the stove, which contains the flame of the stove and also tunnels the flame in an upward direction.

b. Flame Reflector. This allows the heat that would be lost to be reflected back into the heat source.

c. Jet. This shoots a small stream of fuel up to the burner head.

d. Priming Cup. This holds fuel that will be used to prime the stove and also helps to thaw the stove.

e. Generator Tube Assembly. This runs up through the flame reflector to the burner head and helps in vaporizing the fuel.

f. Fuel Tube Assembly. This is the link between the stove and the external fuel bottle.

g. Alignment Block. This is attached to the end of the fuel tube assembly and keeps it aligned when inserted into the pump assembly.

h. Pump Assembly. This is inserted into the fuel bottle to pressurize the fuel for flow of fuel to the stove.

i. Fuel Bottle. This contains the fuel to be burned in the stove.

8. **SERVICEABILITY CHECK FOR THE MSR WHISPERLITE STOVE**.

a. Burner Head. Not loose or bent, which could result in a poorly functioning, stove.

b. Flame Reflector. Not missing or bent out of shape.

c. Jet. Make sure it is in place and not clogged up with carbon or debris from the fuel bottle.

d. Priming Cup. Ensure that the cup has no holes in which the fuel will leak out of and is not loose.

e. Generator Tube Assembly.

(1) There should not be any kinks or sharp bends in the tube.

(2) The tube should not contain any leaks.

f. Fuel Tube Assembly.

(1) Check for leaks and kinks in the tube.

(2) It should be permanently attached to the generator tube assembly.

g. Alignment Block. The alignment block should be attached to the end of the fuel tube assembly and fit snugly to the pump assembly.

h. Pump Assembly.

(1) The majority of the pump assembly is made from plastic, extreme cold temperatures could cause the plastic to become brittle so check for cracks.

(2) The leather cup at the end of the pump assembly should be lubricated in order for it to pressurize the fuel bottle.

(3) The seal around the pump assembly should be pliable and capable of retaining the fuel within the fuel bottle.

i. Fuel Bottle.

(1) Check for leaks and condition of the threads in which the fuel pump assembly is attached to.

(2) Ensure that no debris is inside the fuel bottle, which could possibly clog the fuel line.

9. **LIGHTING THE MSR WHISPERLITE STOVE**.

a. Ensure that the fuel tank has sufficient, *clean* fuel in it. No more than 3/4 full to allow for pressurization.

b. Connect the fuel tube assembly to the pump assembly and place the catch arm over the top of the fuel control valve.

c. Pump only enough pressure into fuel bottle.

d. Turn the fuel control valve on until there is sufficient amount of fuel in the priming cup and then shut the fuel off.

e. Light the fuel contained within the priming cup and let burn until the flame is nearly out.

f. Immediately turn the fuel on slowly until you get a steady blue flame from the burner head. If the flame is yellow in color, turn the valve down and allow more time for the stove to preheat. The stove is ready to be used.

***NOTE***: Repriming the stove may be necessary under severe cold conditions or if the stove is frozen.

10. **NOMENCLATURE OF THE PEAK 1 STOVE**. The Coleman Peak 1 is a multi-fuel stove. It can burn white gas or kerosene. It is important to be able to identify the different parts. The list corresponds to the numbered diagram on the following page:

(1) Fount.

(2) Filler Cap.

(3) Valve Assembly.

(4) Burner Bowl.

(5) Burner Ring Set.

(6) Pump Cup.

(7) Pump Plunger.

(8) Air Stem and Check Valve.

(9) Clip for Pump Cap.

(10) Burner Box Assembly.

(11) Screw (Six)

(12) Generator Bracket.

(13) Grate.

(14) Generator Assembly.

(15) Lanyard.

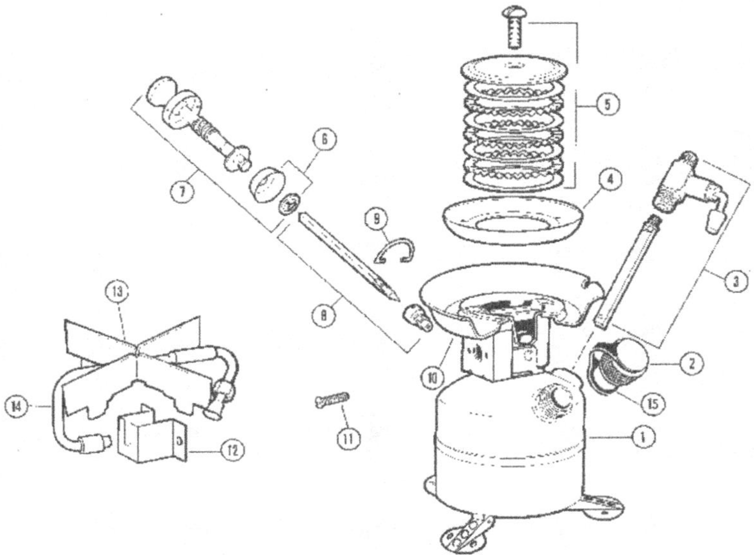


PEAK 1 STOVE

11. **SERVICEABILITY CHECK FOR THE PEAK 1 STOVE**.

a. Burner Assembly. Make sure the screw is tightened down.

b. Grate. Ensure that this is not loose, bent or damaged.

c. Pump Cup. Make sure that this is not bent and seats well into the fuel tank. Lubrication may be necessary to provide efficient pressure.

d. Pump Plunger. Check all the parts for cracks.

e. Filler Cap. Ensure that the filler cap has a gasket and a tight fit to the fuel tank.

f. Pump Cap Clip. Ensure that this is in place to hold the pump assembly stable while pressurizing the fuel tank.

g. Generator Assembly. Check for kinks or fuel leaks.

h. Valve Assembly. Ensure the threads are not damaged and that they fit properly into the fuel tank.

i. Fount. Check for fuel leaks and that the proper, clean fuel is used in the fuel tank.

12. **MAINTENANCE FOR THE PEAK 1 STOVE**.

a. Parts. Ensure that all parts of the stove are tight. If this is not done periodically, you will have a stove falling apart on you and fail to operate properly.

b. Carbon Buildup. By taking a small toothbrush and a small amount of fuel, removal of carbon from the following areas will be possible:

(1) Burner head.

(2) Outside area of the burner head.

(3) The entire stove itself due to spilt food.

c. Plastic or Rubber Parts. If at all possible, keep the stove inside a pack or wrapped in clothing and out of the extreme cold until use to prevent possible cracking and damage.

13. **LIGHTING THE PEAK 1 STOVE**.

a. Ensure that the fuel tank has sufficient fuel in it. No more than 3/4 full to allow for pressurization.

b. Place the stove on a level surface. *DO NOT TIP THE STOVE*.

c. Be sure that the control knob is in the "off" position.

d. Open the pump knob one turn counterclockwise.

e. With the thumb over the hole in the pump knob, pump air into the fuel tank, DO NOT OVER PRESSURIZE THE FUEL TANK. If little or no resistance is felt, lubricate or replace the pump cup.

f. Close pump knob firmly to the right.

g. Hold a lit match to the burner bowl.

h. Turn the black control knob counterclockwise to the "HI" position. If a yellow flame or liquid fuel appears in the burner, turn the control knob "OFF" and allow the flame to burn out excess fuel or allow it to evaporate before relighting.

i. It may be necessary to repump the stove occasionally during use for full heat output.

j. To regulate the heat, turn the control knob between "HI" and "LOW".

k. To turn the stove off, turn the control knob fully clockwise to the "OFF" position and the flame will slowly extinguish itself.

**UNITED STATES MARINE CORPS**
Mountain Warfare Training Center
Bridgeport, California 93517-5001

LP
10/23/01

## STUDENT HANDOUT

## TEN-MAN ARCTIC TENT

**PURPOSE**. The purpose of this period of instruction is to familiarize the student with the 10 man arctic tent, its nomenclature, and its set-up and take down procedures. This lesson relates to bivouac routine.

**OUTLINE.**

1. **NOMENCLATURE AND SERVICEABILITY**.

   a. Apex. This is the metal plate found in the middle of the tent.

   b. Stove Hole. This is the hole on the side of the tent with a rubber ring around it. The rubber should be serviceable and the covers for the hole should be rolled up and tied off when a stove is used.

   c. Front Door. The front door is the door nearest the stove hole. All zippers should work and all buttons should be present.

   d. Back Door. The back door is opposite the front door and should always be secured. Ensure that the zippers work and that the buttons are present.

   e. Primary Lines. These are gold colored lines on the vertical seams of the tent. There should be one on each vertical seam.

   f. Intermediate Lines. These are the white lines on the horizontal seam between the primary lines. There should be a line between each of the two primary lines.

   g. Secondary Lines. Secondary lines are white lines located above the primary lines on the vertical seams. There should be one secondary line above each primary line.

   h. Canvas. The canvas should be free of holes. Any holes should be patched up or sewn up.

i. Tent Cover. This should also be free of holes and should have straps in order to secure the folded tent.

j. Telescopic Pole. The tip of the telescopic pole must fit into the apex and be able to extend without collapsing when the weight of the tent is placed on it.

k. Liner. The tent liner is white and should be free of holes. All holes must be sewn or patched. The liner doors must have zippers that work.

2. **SITE SELECTION**. There are a few basic criteria that should be considered in choosing a site to erect the ten-man arctic tent. Forested areas offer the best site. This area provides cover and concealment for the Marines. The trees also provide protection from the wind. Wooded areas also provide firewood and other materials for construction of defensive positions. If fires are built, the tree overhead will help disperse the smoke. These are just a few principles in the selection of a site. For more guidance on site selection refer to Bivouac Routine.

3. **ERECTION**. In order to set up the ten-man tent, a level area must be created. If the snow is deep, it can be shoveled out or packed down. The diameter of the tent can be measured by using a ski pole held at arms length. Two men will be needed; each man will hold one end of the ski pole. One man will position himself in the center of the prospected site for the tent; the other man will walk a complete circle, which as a result, will determine the diameter needed. The following steps are taken in order to erect the ten-man tent:

a. The tent should be laid out in a hexagonal shape, with the front door, the end with the stove opening, 11 or 1 o'clock, with the wind to your back, downwind, canvas side up.

b. Primary Lines. Lay out the gold primary lines down the seams of the tent.

(1) Stake down. Take the tent pins and prepare to stake down these lines or tie these lines to the trees that are close enough to the tent for anchoring.

c. Telescopic pole. One man with the telescopic pole extended, opens the door, enters the tent and sets the pole in the ring at the peak of the tent; he then raises the pole to a vertical position.

(1) Do not twist the pole or it may collapse.

(2) It is advisable to place a hard object such as a piece of wood or an MRE box under the tent pole to keep it from sinking into the snow or ground.

d. Primary Lines. Once the tent is raised, the primary lines can be secured and tightened.

(1) Driving the tent pins, or "dead men", in about 7 feet away, in line with the seams does this.

(2) Secure opposite lines to ensure an even pull.

e. Stake down the corners and doors.

f. Secondary Lines. Once completed, the secondary lines can be staked out. Do this about one foot further out than the corner stakes.

g. Intermediate lines can be staked down and tightened.

h. Securing the Tent. All the tent lines can be adjusted to ensure a tight and secure tent.

i. Bottom Flap. Now the ten-man tent is completed except for the bottom flap. Tuck the flap under and cover it with the occupant's gear in wet cold conditions.

(1) In extreme dry cold, where the snow is dry and ice will not form, the flap can be folded out and covered with snow.

j. Tents Married Together. Several tents can be married together with the zippers to form CP's or BAS's.

4. **DEAD MAN**. The "dead man" is used when no natural anchors such as trees, logs, branches, or stumps are available.

a. In Deep Snow. This is accomplished by digging a hole large enough to insert a pole or log approximately 3 feet long with the tent line attached. The hole is then filled with well-packed snow, and in a short period of time the packed snow will freeze, and the tent will be secured.

b. On Ice. When the thickness of the ice is not excessive, a small hole is chopped through the ice.

(1) A short stick or pole with a piece of rope or wire tied in the middle of it is pushed through and then turned across the hole underneath the ice.

(2) If the ice is very thick, a hole 1 to 2 feet deep is cut in it. The "dead man" is then inserted in the hole and filled with slush or water.

(3) When the slush or water is frozen, an excellent anchor point is provided.

(4) When the "dead man" is placed underneath or into the ice, a piece of rope or wire should be fastened to the rope or wire after the "dead man" is secure. This may prevent the tent line from being accidentally cut or damaged when being removed from the ice.

5. **STRIKING**. Many times it will be necessary to brush off any excessive snow or frost from both the inside and the outside of the tent. Once this is done, the following steps are to be taken in order to strike the ten-man arctic tent:

a. Remove the tent pins/stakes holding the tent lines down.

b. Once the tent is laid out flat, close the door zippers and snaps, and tie up all lines securely.

c. Now the tent is ready to be folded.

(1) Lift the peak and one corner of the tent off the ground. Then bring the next corner.

(2) Repeat placing the corners together until the panels have been folded into an accordion pleat.

(3) While bringing the corners together, make sure the inside liner is folded smoothly alongside the tent.

(4) Lay the tent down, even out the folds, and place the tent lines into the tent.

(5) Fold the top half down to the eave, and then fold the bottom up and over the eave.

d. Place folded tent in cover and strap-up tightly.

e. Storage and Movement. Now the ten-man arctic tent is ready for storage or movement in the ahkio.

**UNITED STATES MARINE CORPS**
Mountain Warfare Training Center
Bridgeport, California 93517-5001

FMST.07.37
10/23/01

**STUDENT HANDOUT**

**BURN MANAGEMENT**

**TERMINAL LEARNING OBJECTIVE**. Given a simulated casualty, treat burn injury casualties, in accordance with the references. (FMST.07.37)

**ENABLING LEARNING OBJECTIVES**

(1). Without the aid of references, given a list of the depths and appearances of burns, match them to the type of burn, in accordance with the references. (FMST.07.37a)

(2). Without the aid of references, given a list, choose the major burns that require rapid casualty evacuation, in accordance with the references. (FMST.07.37b)

(3). Without the aid of references, given a list, choose the first step in the treatment of burn injuries, in accordance with the references. (FMST.07.37c)

**OUTLINE**

1. **GENERAL:** Each year there are about 2 million individuals burned severely enough to seek medical attention. Around 70,000 require hospitalization. Deaths are on the decrease largely due to the increase in the use of smoke detectors. It has been estimated that 90% of all burn injuries are preventable. Children and young adults are frequently affected by burn injuries.

2. **PHYSIOLOGY:** For non-chemical burns, the primary events of injury occur during the time of heat contact. Coagulation necrosis takes place within cells and denaturation of collagen in the dermis. Blood vessels are completely destroyed or endothelium damage is severe enough to cause clotting which leads to ischemic necrosis of remaining viable cells. Burn wounds are not static. Surrounding the "coagulation zone" is a zone of capillary and small vessel stasis. Circulation becomes stagnant from a strip formed by red blood cells, and aggregation of platelets and white blood cells. The fate of the burn wound depends on the progression of this zone of stasis. Cells release mediators to start the inflammatory response.

**18**

3. In patients with burns less than 10% of the total body surface area (TBSA), the mediators are generally limited to the burn site itself. As burns approach 20% TBSA, the local response

becomes systemic. Burns of 30% TBSA or greater, often lead to acute renal failure, which in a patient with a severe burn almost invariably leads to a fatal outcome.

Cardiovascular changes begin immediately after a burn. The degree of these changes depends initially on the size of the burn and to a lesser extent the depth of the burn. Burns less than 15% TBSA can be fluid resuscitated with crystalloid fluids. As burns pass 20% TBSA, massive shifts of fluid and electrolytes occur from intravascular into extracellular space. Unless intravascular volume is repleted, classic hypovolemic shock occurs.

3. **TYPES OF BURNS**.

a. Scald Burns: Scalding is the most common cause of burns, usually resulting from hot water. Water at 140°F creates a deep partial-thickness or full-thickness burn in three seconds. At 156°F the same burn occurs in one second. Freshly brewed coffee is generally about 180°F. Boiling water always causes a deep burn. Soups and sauces are thicker, remaining in contact with the skin longer. Exposed areas tend to burn less deeply than areas covered with a thin layer of clothes. Clothing retains heat and keeps liquid in contact with the skin longer. Immersion scalds are deep and cause severe burns resulting from longer contact time. These burns are more common in children and elderly patients because of their thin skin. Burns from hot oils are generally deep partial-thickness or full-thickness. Temperatures for cooking oils reach around 400°F, and tar can be up to 500°F.

b. Flame Burns: The next most common burn injuries are flame burns. Although household injuries have gone down with the advent of smoke detectors, a high number of burn injuries still occur from careless smoking, improper use of flammable liquids, automobile accidents, and clothing ignited from stoves or space heaters.

c. Flash Burns: Flash burns result from intense heat for a brief time caused by explosions of natural gas, propane, gasoline, and other flammable liquids.

d. Contact Burns: These burns usually result from direct contact of hot metals, plastics, glass, or coals. Contact burns are usually limited in size, but are deep wounds. With the increase of wood stove usage, this type of burn injury is on the rise in children. Contact burns, dealing with an unconscious patient or involving molten materials, are usually fourth degree. The most common contact burn in the wilderness is from hot coals.

e. Electrical Burns: Electrical burns are actually thermal burns from very high-intensity heat. When electricity encounters resistance from the body it turns into heat. The severity of the burn depends on the portion of the body affected to amperage ratio. The smaller the body part, the more intense the heat is, and the part is less able to dissipate the heat. Consequently the fingers and toes are almost always destroyed with severe damage to the foreams. Larger areas can dissipate the current enough to prevent extensive damage. Arc burns take the most direct route, not the path of least resistance. These are deep and destuctive burns which mostly occur at the joints.

Electrical burns may have other associated injuries that should not be overlooked on the initial assessment. Fall related injuries are common. The intense muscle contraction resulting from electrocution can cause fractures of the lumbar vertebrae, humerus, or femur, and also may cause dislocation of the hips and shoulders. The casualty may have cardiac symptoms of a myocardial contusion or infarction. There can be a conduction system failure or an actual rupture of the heart wall. If there are no signs present of cardiac problems intially after a shock of 110 or 220 volts there is a minute chance they will appear later. Nervous system damage is possible whenever a current passes from one side of the body to the other more than likely affecting the spinal cord.

f. Chemical Burns: Usually caused by strong acids or alkalis and are most often the result of industrial accidents, home use of drain cleaners, assault, and improper use of harsh solvents. Chemical burns continue to burn until they are inactivated by reaction with tissue or dilution by flushing with water. Acids tend to tan the skin, creating an impermeable barrier that limits futher penetration. Alkalis combine with cutaneous lipids and saponify[1] the skin until they are neutralized. The appearance of the wound can be misleading. Unless the observer can be 100% sure, all chemical burns should be considered full thickness.

g. NBC Burns: Nuclear weapons cause burns in two ways: by direct absorption of thermal radiation through exposed surfaces (flash burns); or by the indirect action of fires caused by the weapon (flash burns). The first step in treatment is to prevent contamination to the rescuers and other patients.Vesicants such as mustard gas cause skin burns with edema and blister formation. Burns resulting from vesicants should be flushed with copious amounts of water. Absorbent powder (flour, talcum powder, Fuller's earth) can be used if water is scarce. The powders should be wiped away with a moist towel. The military uses M258A1 kits for skin decontamination. White Phosphorus ignites spontaneously when exposed to air. Remove the air with an air-tight seal; mud is optional. A brief rinse with 1% copper sulfate[2] will impead further oxidation.

4. **CLINICAL PRESENTATION**: The severity of the burn injury is related to the size of the burn, the depth of the burn, and the part of the body that is burned.

a. Burn Size: Burns are the only quantifiable form of trauma. The burn size is the single most important factor in predicting mortality, need for specialized care, and the complications expected from the burn. We can measure burn size with the "rule of nines". The "rule of nines" is where areas of the body represent approximately 9% of a person's total body surface area (TBSA), each leg 18% (the front of the leg 9%, and the back of the leg 9%), the front of the torso represents 18%, the back of the trunk is 18%, the head represents 9%, and the perineum 1%. For infants and small children under four years of age, the head represents a larger percentage 18%, and the legs a smaller percentage 13%, each arm 9%, front troso 13%, back troso 13%, and the perineum 1%. The percentage of the burn area and that of the unburned area need to add up to equal 100%. For smaller burn areas, use the Rule of Palmar Surface: the patient's palmar surface equals about 1% TBSA.

---

[1] Saponify: to convert fat into soap by treating it with an alkali.
[2] Navy uses sodium bicarbonate instead of copper sulfate

b. Burn Depth: The depth of the burn is either described in degree: first, second, third or fourth, or by the depth of the injury. The following describes burns in terms of partial-thickness or full-thickness. While these descriptions appear to separate burns into defined categories, many burns have a mixture of characteristics making a precise diagnosis difficult.

(1) First-Degree: These burns involve the epidermis only and do not blister. They are most commonly caused by ultraviolet light. An example would be a sunburn. The burned skin is painful and red. It should heal in 7 days without scarring.

(2) Superficial Second-Degree, or Superficial Partial-Thickness: These burns include the epidermis and upper layers of the dermis. They characteristically form blisters, under which the skin is red and moist, these burns are painful to the touch. Blisters can take up to two days to appear. These burns are usually caused by hot liquids. Wounds should heal in 14 to 21 days; they may scar depending on the extent of the burn.

(3) Deep Second-Degree, or Deep Partial-Thickness: Burns that involve deeper layers of the dermis. There is damage to hair follicles and sweat glands. They are characterized with a molted pink and white color with blisters forming immediately. These wounds may be less sensitive to touch than the surrounding normal skin, or maybe tender to touch. The patient often complains of discomfort rather than pain. Capillary refill may be slow or absent, when pressure is applied to the wound. These burns are caused by hot liquids, oil, steam, or flame. They may be difficult to distinguish from Third-Degree/Full-Thickness burns. Healing takes 3-9 weeks. Scarring is probable, the degree is related to the amount and the depth of dermal injury. Surgical grafting may be required.

(4) Third-Degree, or Full-Thickness: Involves the entire thickness of the skin, epidermis through the dermis, down to subcutaneous fat. All structures of the epidermis and dermis are destroyed. The wound is classically described as leathery, firm, and depressed when compared to normal skin. The tissue is charred, pale, and insensitive to light and touch. It is often misdiagnosed as a Deep Second-Degree/Deep Partial-Thickness due to the similar clinical findings. This burn will not spontaneously heal. Surgical repair and/or skin grafts are necessary. There is significant scarring.

(5) Fourth-Degree: These burns involve all layers of the skin, subcutaneous fat, muscle, and bone. These are devastating, life threatening injuries. They almost always have a charred appearance, and often only the cause of the burn gives a clue to the amount of underlying tissue destruction. (FMST.07.37a)

c. Burn Classification: The American Burn Association has devised a classification of burns, dividing them into major, moderate, and minor burns. Patients are placed in groups related to their risk. Low-risk patients are between the ages of 10 and 50 years old. High-risk patients are less than 10 and greater than 50 years old. Poor-risk are patients with underlying medical illnesses such as heart disease, diabetes, or chronic pulmonary problems.

(1) Minor Burns: Imply outpatient treatment. Minor burns involve TBSA of less than 15% in the low-risk group, or less than 10% TBSA in the high-risk group; full-thickness burns of less than 2% TBSA in anyone; and no other injuries.

(2) Moderate Burns: Are partial-thickness burns of 15-25% TBSA in the low-risk group, 10-20% TBSA in the high-risk group; full-thickness burns of equal to or less than 10% TBSA in anyone; burns not involving the hands, face, feet, perineum, or circumferential limbs.

(3) Major Burns: Defined as partial-thickness burns greater than 25% TBSA in the low-risk group, or greater than 20% TBSA in the high risk group; full-thickness burns of greater than 10% TBSA in anyone; burns involving the hands, face, feet, or perineum; burns crossing major joints; circumferential limb burns; burns complicated by inhalation injury; electrical burns; burns complicated by fractures or other forms of trauma; and burns in poor-risk patients. These burns require rapid casualty evacuation. (FMST.07.37b)

5. **TREATMENT:**

a. Initial Care:

(1) The first step in the treatment of burn injuries is to stop the burning process. (FMST.07.37c) Remove patients from heat, remove all their clothes that are affected by hot liquids, shut off electricity (very important), remove chemicals by flushing the affected area for 10 to 15 minutes, and decontamination of NBC. Heat can continue to injure tissue for a surprisingly long time. No first aid will be effective until the burning process has stopped.

(2) Manage the ABC's. Once an airway has been established give oxygen. If qualified, consider intubation earlier rather than later.

(3) Assess for associated injuries such as fractures or lacerations and inhalation injury.

(4) Remove clothing and jewelry from the burn area. This will help prevent constriction from swelling of the burn tissue. Do not try to remove anything that has adhered to the wound.

(5) Evaluate the Burn (depth, extent, pain).

b. General Treatment for the Patient:

(1) Gently wash the burn with tepid water and mild soap to remove any debris and to clean the skin surface around the burn site. Pat dry. Remove the skin from blisters that have popped open (do not open blisters unless necessary for function of hands or feet).

(2) Dress burn with a thin layer of antibiotic ointment.

(3) Cover the burn with Second Skin (if the burn is small enough) or cover with a thin layer of gauze, or clean, dry clothing. Burns on the face, neck and hands may be left open to the air after applying silvadene ointment. Covering the wound reduces pain and evaporative losses, but do not use an occlusive dressing.

(4) When evacuation is imminent, do not redress or reexamine the injury. If evacuation is prolonged, redress once a day. Remove old dressings, reclean (removing the old ointment), and apply fresh ointment and a clean dry covering. (Note: if stuck, soak off old dressings with clean, tepid water.)

(5) Do not pack wounds, which are larger than 20% of the body surface area, in ice. Do not leave wet coverings on burns for more than two hours at a time to reduce the risk of hypothermia.

(6) Stabilize the body temperature. When skin is lost, so is the patient's ability to thermoregulate.

(7) Have the patient drink as much fluid as he or she can tolerate without vomiting. Include some salt in the oral fluids, but do not make these solutions stronger than . 9%. This is the equivalent of a pinch of salt per 8 ounce glass.

(8) IV therapy as follows, Baxters (Parkland) Formula: In the first 24 hours give Ringers Lactate, 4ml times the weight in kilograms times the percent TBSA burned. Half of the solution is given in the first eight hours. The second half is given over the next 16 hours. The second 24 hours should be given fluids to maintain blood pressure, colloids (usually in the form of albumin) are the preferred choice.

4 x wt (kg) x TBSA= amount LR given in the first 24 hours

Example a 70kg patient with 50% TBSA this is the formula for IV therapy for the first 24 hours.

4 x 70kg x 50% = 14,000ml in 1ST 24 hours
7,000ml hours 1-8
3,500ml hours 9-16
3,500ml hours 17-24

c. General Management of the Burn: Caring for the wound itself is often the least important aspect of burn care. All burn wounds are sterile for the first 24 to 48 hours. Burn management is aimed at keeping the wound clean and reducing the pain.

(1) Elevate burned extremities to minimize swelling. Swelling retards healing and encourages infection. Have the patient gently and regularly move burned areas as much as possible.

(2) Burns are tetanus-prone wounds. Check on the last known tetanus shot. The patient may need a renewal.

(3) Ibuprofen is probably the best over the counter analgesic for burn pain (including sunburn).

(4) If you have no ointment or dressings, leave the burn alone. The burn's surface will dry into a scab like covering that provides a significant amount of protection.

6. **INHALATION INJURIES:** Half of all fire-related deaths are from smoke inhalation. Smoke inhalation doubles the mortality rate for any burn. Smoke inhalation is a general term for carbon monoxide poisoning, thermal airway injury, and smoke poisoning.

a. Carbon Monoxide Poisoning: Carbon monoxide (CO) is a colorless, odorless, tasteless gas that has an affinity for hemoglobin 200 times greater than that of oxygen. Carboxyhemoglobin (COHb) levels are measured as the percentage of hemoglobin bound to carbon monoxide. The following COHb levels correlate with clinical symptoms:

| | |
|---|---|
| <10% | No symptoms |
| 20% | Headache, nausea, vomiting, loss of dexterity |
| 30% | Confusion, lethargy, ECG ST depression |
| 40% | Coma |
| 50% | Death |

The half-life of COHb is 4 to 5 hours at room air, 90 minutes with 100% oxygen, and 20 to 25 minutes at 3 atm in a hyperbaric chamber. Treatment for less than 40% COHb is a 100% oxygen. At 40% or greater the hyperbaric chamber should be used. If patient is still in a coma after COHb levels are normal, prognosis is poor, and it has been experience that they rarely awaken.

b. Thermal Airway Injury: Injury to the respiratory tract from inhalation of hot gas or steam. The lower respiratory tract is usually not damaged, unless there was an inhalation of flame or steam. Any patients who were in an explosion, with burns of the hands, face, and upper torso are at risk. Signs of carbon such as soot in the pharnyx should alert the caregiver to airway injury. Maintaining the airway is the important issue here.

c. Smoke Poisoning: Is the inhalation of noxious gases that are the products of combustion. Cyanide poisoning is known to occur in victims of smoke inhalation. Other toxic gases released are sulfur dioxide, hydrogen chloride, phosgene, and ammonia. These gases cause damage to the respiratory tract similar to chemical burns. Physical signs include facial burns, intraoral or pharyngeal burns, singed nasal hairs, soot in mouth or nose, hoarseness, carbonaceous sputum, and expiratory wheezing.

**UNITED STATES MARINE CORPS**
Mountain Warfare Training Center
Bridgeport, California 93517-5001

FMST.07.10
10/23/01

**STUDENT HANDOUT**

# MISCELLANEOUS COLD WEATHER MEDICAL PROBLEMS

**TERMINAL LEARNING OBJECTIVE.** Given an actual/simulated casualty in a cold weather/mountainous environment, treat miscellaneous health problems, in accordance with the references. (FMST.07.10)

**ENABLING LEARNING OBJECTIVES.**

1) Given an actual/simulated casualty in a cold weather/mountainous environment, treat snow blindness, in accordance with the references. (FMST.07.10m)

2) Given an actual/simulated casualty in a cold weather/mountainous environment, treat dehydration, in accordance with the references. (FMST.07.10n)

3) Given an actual/simulated casualty in a cold weather/mountainous environment, treat carbon monoxide poisoning, in accordance with the references. (FMST.07.10o)

4) Given an actual/simulated casualty in a cold weather/mountainous environment, treat corneal frostbite, in accordance with the references. (FMST.07.10p)

5) Without the aid of references, select from a given list the correct definition of snow blindness, in accordance with the references. (FMST.07.10q)

6) Without the aid of references, select from a given list the correct signs and symptoms of snow blindness, in accordance with the references. (FMST.07.10r)

7) Without the aid of references, select from a given list the correct treatment(s) of snow blindness, in accordance with the references. (FMST.07.10s)

8) Without the aid of references, select from a given list the correct definition of dehydration, in accordance with the references. (FMST.07.10t)

**19**

9) Without the aid of references, select from a given list the correct signs and symptoms of dehydration, in accordance with the references. (FMST.07.10u)

10) Without the aid of references, select from a given list the correct treatment of dehydration, in accordance with the references. (FMST.07.10v)

11) Without the aid of references, select from a given list the correct definition of carbon monoxide poisoning, in accordance with the references. (FMST.07.10w)

12) Without the aid of references, select from a given list the correct signs and symptoms of carbon monoxide poisoning, in accordance with the references. (FMST.07.10x)

13) Without the aid of references, select from a given list, the correct treatment(s) of carbon monoxide poisoning, in accordance with the references. (FMST.07.10y)

**OUTLINE.**

1. **SNOW BLINDNESS.**

a. Definition. Snow blindness is burning of the conjunctiva and superficial cells of the cornea by ultraviolet light. (CWM.6.8e)

b. General. Snow blindness is caused by ultraviolet light of 200-300 nm wavelengths. Symptoms of snow blindness usually occur 2-12 hours after exposure. Like sunburn, the danger of snow blindness is greatest not on a clear, bright day but rather on a dull, cloudy day or during a whiteout.

c. Signs/Symptoms. (CWM.6.8f) Snow blindness symptoms usually last one to five days depending on the severity of exposure, but some patients have noted symptoms for as long as five to seven days. These symptoms consist of:

(1) Gritty sensation in the eyes.

(2) Pain.

(3) Increased tearing.

(4) Hot sensation.

(5) Photophobia.

(6) Blurred vision.

(7) Headache.

(8) Swollen eyelids.

d. Treatment. (CWM.6.8g) The patient with snow blindness should have an eye exam, which includes flourescin staining of the conjunctiva. In severe cases, one sees punctuate uptake of flourescin stain over the cornea. The patient should be treated with antibiotic ointment and patching both eyes. Occasionally, aspirin, Tylenol, codeine, or even strong narcotics may be needed for the headache/pain.

(1) It is important not to use local anesthetics, i.e., Alcaine, because these inhibit regeneration of corneal epithelium. Also, do not use steroid drops as these may facilitate the development of infection.

(2) Continue to re-examine the eye on a daily basis until it is healed. Mild cases usually recover in 24-48 hours. Severe cases may take up to three to four days to recover. Once a person has suffered snow blindness he is more prone to future recurrences.

e. Prevention. Prevention is critical, because this injury is totally preventable if troops wear proper eye protection. Almost any sunglasses, if worn, will prevent snow blindness. Sunglasses made of G-15 glass are excellent. For extreme situations G-15 glass coated with a nickel metallic coating (double gradient) provides the best protection.

(1) If one is in a survival situation without sunglasses, then slit goggles can be fashioned Eskimo style.

2. **CALOROPHTHALGIA**.

a. Definition. Calorophthalgia is the sensation of conjunctival heat and eye discomfort caused by excessive exposure of the eyes to infrared light.

b. Signs/Symptoms. The patient with calorophthalgia will present with the sensation of conjunctival heat and indescribable discomfort deep in the eye. This discomfort creates the desire to turn the head, to close the lids, to further protect the eyes by forcing the brows and cheeks over the orbits, and, if severe, to put the hands over the orbits and press for a time until the discomfort ceases.

(1) Calorophthalgia is caused by excessive infrared transmittance by sunglasses. The calorophthalgic index (see references by E. E. Hedblom) will predict whether or not a pair of sunglasses will prevent calorophthalgia.

c. Treatment. Antibiotic (Blephimide) drops every 2 hours, while awake, for two days.

d. Preventive Measures. Consists of providing the individual with an adequate pair of sunglasses.

3. **DEHYDRATION**.

a. Definition. (CWM.6.8h) Dehydration, more correctly termed volume depletion, is defined as a deficit of total body water.

b. General. Dehydration is probably the greatest enemy of troops in cold weather operations. Troops at altitudes will suffer larger than normal fluid losses due to increased respiratory rate of dry air. Additionally, the cold itself will induce a diuresis and since potable water is usually not readily available (usually snow must be melted) the tendency not to drink adequate water is compounded. The result is a chronic state of volume depletion. America's favorite hot drinks include coffee and tea, both of which contain caffeine, a diuretic. Drinking hot fruit cups, as is done in Scandinavia, would be more appropriate.

**NOTE**: The British in the Falkland war found that volume depletion was rampant. Consequently, when a soldier was wounded, a smaller than usual blood loss was apt to send him into shock. This was attributed to a contracted blood volume due to chronic volume depletion.

c. Signs/Symptoms. (CWM.6.8i) The signs and symptoms of volume depletion are headache, dizziness, dry mucous membranes, dark urine, increased pulse, constipation, irritability, fatigue, sleeplessness, abdominal pain, etc. Volume depletion predisposes to flaring up of hemorrhoids (the number one medicine prescribed by the British in the Falklands was hemorrhoid cream).

d. Treatment. (CWM.6.8j) The treatment of choice is fluid replacement. Mild cases can be treated with PO fluids. Severe cases will require IV replacement of fluid loss.

e. Prevention becomes extremely important. Troops at altitude should be forced to drink at least 6 quarts of water each day.

4. **CARBON MONOXIDE POISONING**.

a. Definition. (CWM.6.8k) The definition of carbon monoxide poisoning is the inhalation of carbon monoxide to a degree that symptoms occur.

b. Carbon monoxide (CO) is a heavy, odorless, colorless, tasteless gas resulting from incomplete combustion of fuels. CO kills through asphyxia even in the presence of adequate oxygen because oxygen transporting RBC hemoglobin has a 210 times greater affinity for CO than for oxygen.

(1) Recent research indicated that the pathophysiology of CO poisoning may be more complex than this and that there is a dismal lack of information concerning the detailed biochemical mechanisms involved (see references).

c. At sea level, carboxyhemoglobin concentrations of up to 10% cause no symptoms (heavy smokers run up to 8% levels). At carboxyhemoglobin concentrations of 20%, symptoms of mild poisoning occur. At carboxyhemoglobin concentrations of 30-50%, signs and symptoms of severe carbon monoxide poisoning occur. Carboxyhemoglobin concentrations greater than 50% are generally fatal.

d. Signs/Symptoms. (CWM.6.8l) The signs and symptoms depend on the amount of CO the victim has inhaled. In mild cases, the victim may have only dizziness, headache, and confusion; severe cases can cause a deep coma. Sudden respiratory arrest may occur. The classic signs of CO poisoning are cherry-red skin color; in practice, the skin may be pale, blue, or red.

(1) CO poisoning should be suspected whenever a person in a poorly ventilated area suddenly collapses. Recognizing these conditions may be difficult when all members of the party are affected.

e. Treatment. (CWM.6.8m) The first step is to immediately remove the victim from the contaminated area.

(1) Victims with mild CO poisoning who have not lost consciousness need fresh air and bed rest for a minimum of four hours. More severely affected victims may require rescue breathing.

(2) In sub acute cases, a mental status examination (standard police sobriety test) may be the best diagnostic indicator when the victim is initially seen.

(3) Fortunately, the lungs excrete CO within a few hours.

**NOTE**: Recurrent symptoms can occur in non-pressure chamber treated victims, respond favorably to pressure chamber treatment days after the initial poisoning.

f. Prevention. Prevention is the key. Ensure that there is adequate ventilation when running vehicle engines or when cooking over open flames.

**SPECIAL NOTE**. One of the three rationales for using Hyperbaric Oxygen is that it increases the rate of elimination of CO. The rate has been shown to be 320 minutes when breathing fresh air, 80 minutes when breathing 100% oxygen via mask and dropping to as little as 23 minutes during 100% oxygen at three atmospheres (approximately 66 feet).

5. **CORNEAL FROSTBITE**.

a. Definition. Actual freezing of the cornea. Occurs when large amounts of air pass against and cool the cornea, corneal clouding and loss of vision occur. Commonly occurs in snowmobile drivers not wearing proper eye protection.

b. Signs/Symptoms. On examination the cornea will be clouded over.

c. Treatment and Prevention. The only treatment available is corneal transplantation. Prevention is the key.

6. **HEAT EXHAUSTION**. Contrary to expectations, heat exhaustion is fairly common in cold weather operations. Troops unacclimatized to cold weather tend to overdress and end up overheating and sweating. The logical remedy is to dress so that one is always comfortably cool. This requires removal of clothing layers as you warm up.

7. **COLD HEMOPTYSIS**.

   a. During hyperventilation at low temperatures (less than -25 degrees F) an individual can cough up blood from the tracheobronchial tree. This is not "frostbite" of the tracheobronchial tree as there is no freezing of tissue. Rather, marked respiratory mucosal hyperemia occurs in these situations and lead to the expectoration of blood.

   b. The treatment is reassurance of the patient.

8. **LIGHTNING INJURY**.

   a. High-voltage electrical current from lightning during electrical storms represents a significant hazard in the outdoors. You need to know how to protect yourself from this danger and how to care for a victim of lightning injury.

   b. Lightning is caused by violent vertical air currents associated with the development of cumulonimbus clouds or thunderheads, billowing, vertical clouds with anvil-shaped tops that may tower to 60,000 feet or more. The air currents produce differences of electrical potential between clouds or between a cloud and the earth. Cumulonimbus clouds usually produce large raindrops, huge snowflakes or hail, and tend to develop during the afternoon in warm or hot weather; they also can be part of an advancing cold front.

   c. Lightning causes injury in the same way as any other electric current, except that the duration of the bolt is so short (.0001 to .001 second) that burns are less severe. A person struck by lightning may suffer a characteristic type of superficial skin burn that has a pattern resembling a fern leaf. In addition to direct strikes, ground currents may injure people and side flashes from nearby strikes. Strikes often cause a victim's clothing to explode off the body by instantly converting sweat or other moisture to steam.

   d. Ways in Which Electricity Causes Injury:

      (1) Breathing may stop because of injury to the respiratory control center in the brain.

      (2) Cardiac arrest or ventricular fibrillation can occur because of a direct effect on the heart.

      (3) Paralysis, blindness, numbness, loss of hearing or speech and unconsciousness can occur because of direct effects on the nervous and musculoskeletal systems.

      (4) Severe, deep burns can occur.

(5) Trauma can occur because of falls, frequently caused by strong muscular contractions that throw the victim off balance.

(6) Secondary kidney injury may occur because the broken down products of blood and injured muscle overload the kidneys.

e. Prevention.

(1) If caught in an electrical storm, avoid bodies of water and take shelter away from high points, exposed ridges, and solitary trees. All of these features attract lightning strikes. Return small boats to shore at the first threat of storm. Avoid small caves where body parts are close to cave walls or the ceiling.

(2) If caught in the open, retreat as far down on the side of a ridge or other exposed area as possible and move away from ice axes, ski poles and other metal objects. Squat on your heels until the danger is over. This position shortens the body and minimizes ground contact, decreasing the tendency for the body to act as a lightning rod and making it less likely that ground currents will pass through the body.

(3) During electrical storms at alpine ski areas, ski lifts and exposed summit structures should be cleared of skiers. Avoid metal structures such as lift towers.

f. Field Management.

(1) Perform a rapid and thorough primary survey. (Consider the possibility of a cervical injury.)

(2) Give Basic Life Support (BLS). (Do not give up too soon on CPR or rescue breathing.)

(3) Give oxygen in high concentration, if available.

(4) Take proper precautions if victim is unconscious.

(5) Perform a secondary survey and care for additional injuries, if present.

(6) Monitor and record vital signs at regular intervals.

(7) CASEVAC ASAP!

**UNITED STATES MARINE CORPS**
Mountain Warfare Training Center
Bridgeport, California 93517-5001

FMST.07.17
10/23/01

**STUDENT HANDOUT**

**CASEVAC AND CASEVAC REPORTING**

**TERMINAL LEARNING OBJECTIVE.** Given a casualty in a cold weather environment and necessary equipment and supplies, conduct cold weather casevac procedures, in accordance with the references. (FMST.07.17)

**ENABLING LEARNING OBJECTIVES.**

1) Without the aid of references, list in writing the nine general considerations for a CASEVAC, in accordance with the references. (FMST.07.17a)

2) Without the aid of references, list in writing the five steps in the CASEVAC administrative reporting chain, in accordance with the references. (FMST.07.17b)

**OUTLINE.**

1. **MISSION**. The mission of the Medical Department is to return patients to duty with the Fleet Marine Force, if possible, and to stabilize and prepare for evacuation of those patients whose length of stay is expected to exceed the evacuation policy. The mission is accomplished through the following:

   a. Administration of first aid.

   b. Performance of emergency surgery.

   c. Collection, transportation, sorting, temporary hospitalization and evacuation of casualties.

   d. Technical supervision of measures designed for prevention and control of disease and injury.

   e. Returning personnel to duty within the constraints of the Evacuation Policy.

**20**

2. **GENERAL PRINCIPLES OF MEDICAL SUPPORT**.

a. Principle of Conformity.

(1) Objective: To provide medical support to the sick, injured and wounded personnel at the right time and place.

b. Principle of Proximity.

(1) Objective: To keep injuries and mortality to a minimum by prompt insertion of the casualty into the Medical Support System.

c. Principle of Flexibility.

(1) Objective: To be prepared to shift medical support resources to meet changing requirements.

d. Principle of Mobility.

(1) Objective: To maintain close medical support to maneuvering combat forces.

e. Principle of Continuity.

(1) Objective: To provide optimum care and treatment to the sick, injured and wounded in an uninterrupted manner.

3. **THE MEDICAL EVACUATION CHAIN**.

a. Marine Infantryman. The Marine Infantryman is often overlooked as a member of the treatment team. The Marines are trained in "self-aid" and "buddy-aid". Litter teams are organic to Marine Infantry Companies and are comprised of personnel.

b. Hospital Corpsmen. Hospital Corpsmen are attached to each Marine Company. They provide first aid and emergency procedures, continual observation and care to ensure the airway is open, that bleeding has ceased and shock and further injury is prevented. They also ensure medical supplies are used effectively and make request for air or ground ambulance, as appropriate.

c. Battalion Aid Station. The first link in the chain of evacuation where a patient can expect to see a physician. Treatment is conducted in a relatively safe area, within the battalion's area of operations. Treatment consists of the use of IV fluids, antibiotics, preservation of airway by surgery, if necessary, and the application of more secure splints and/or bandages. For those patients who cannot be returned to duty, the final step is to arrange for the proper means of evacuation.

d. Medical Company/Casualty Receiving and Treatment Ship. The medical company has operating room capability, basic laboratory, pharmacy equipment and supplies, whole blood capability and holding wards. For those patients who require a more comprehensive scope of treatment, arrangements are made for evacuation by ground or air to the particular treatment facility, which can provide the required treatment.

e. Hospital Company. A hospital company is capable of providing principal treatment. It has specialty care and holding beds. Other examples of principal treatment centers are hospital ships, combat zone fleet hospital, rear area zone fleet hospitals and military and civilian hospitals in CONUS.

4. **GENERAL CONSIDERATIONS**. (FMST.07.17a)

a. Apply essential first aid, i.e., splints, pressure bandages.

b. Treat life threatening conditions first, i.e., profuse bleeding, shock, etc.

c. Protect patient from the elements.

d. Avoid unnecessary handling of the patient.

e. Select the easiest route. Scouts may be required.

f. If the route is long and arduous, set up relay points and warming stations using minimal medical personnel.

g. Check on patient's condition frequently.

h. Normal litter teams must be augmented in arduous terrain.

i. Do not separate the patient from his survival gear.

5. **TRANSPORTATION OF THE PATIENT.** There are two types of cold weather mountainous litters used in the military today: standard and improvised.

a. Standard Litter Types:

(1) Ahkio Litter. The standard ahkio can be readily used as a litter. Pad the bottom with sleeping mat and bag and use assistants to prevent the ahkio from overturning. Lash the patient securely, but not too tightly. The patient's head is placed to the rear and slightly elevated unless the evacuation route is downhill, then the head is placed to the front. The ahkio is the preferred litter, because not only can it be manhandled, but pulled by any over-the-snow vehicle ie. Snowmobile, DMC, snow cat, or loaded into a helicopter/aircraft, etc. It is large enough to accommodate a patient and most, if not all, of his survival gear.

(2) Hjelper Sled. A Hjelper sled can be constructed using the injured patients skis and poles. While requiring some time to properly construct, the Hjelper sled is very

energy efficient and can be pulled across the snow with less effort than other litters. Remember not to place the load too far forward.

b. Improvised Litters

(1) Rope Litter *(Clove Hitch Method)*. A rope litter is a field expedient litter prepared using one rope. It requires 20 to 30 minutes to prepare and should be used only when other materials are not available. (Note: Above the tree line very little material exists to construct litters.) May be constructed using 165' coil.

(a) Make 24 bights about 45 centimeters (18 in) long from the rope starting in the middle of the rope. By starting in the middle of the rope two people can work on the litter at once.

(b) With the remainder of the rope make a clove hitch over each bight. Each clove hitch should be approximately six inches apart when the litter is complete.

(c) Pass the remainder of the rope through the bights outside of the clove hitches. Dress the clove hitches down toward the closed end of the bight to secure the litter and tie off the ends of the rope with clove hitches.

(d) Four to six bearers are required for this method.

(e) Depending on the situation, the above measurements may have to be altered slightly.

(f) If padding is available, pad the stretcher.

(g) If ground is snow covered, a trail must be broken.

(2) Poncho Litter.

(a) Spread the open poncho over the ground.

(b) Place one 2.5-meter pole just to the left of the center eyelets and fold the poncho over it.

(c) Place another 2.5-meter pole on the now doubled poncho approximately 30 cm from the first pole.

(d) Fold the poncho over the second pole.

(e) Ensure the ends of the poles extend outside the poncho.

(f) Lay the casualty on the litter.

(g) The body weight of the casualty will hold the litter together.

(h) One litter bearer positions himself at the head of the litter, one at the foot.

(3) Litter Travois.

(a) This litter is made the same way as the shirt/jacket ski litter.

(b) Casualty. Place the casualty's head towards the ski tails and protect him from the elements.

(c) Ski tails. Lift up on the ski tails to allow the ski tips to slide over the snow smoothly.

**NOTE**: For all litters, tape slings worn diagonally across the bearer's shoulders and secured to the litter can free the hands for balance and aid.

6. **EVACUATION BY GROUND VEHICLES**. Vehicle requirements will vary depending on the type of operation and the terrain. Wheeled vehicles are frequently limited to paved roads. Chains are frequently required, even with four-wheel drive. Tracked vehicles are also limited to a degree. Over snow vehicles, i.e., snowmobiles, DMC's and LMC's have limited usage in the moving of patients. A vehicle that solves these problems is the BV-206.

7. **EVACUATION BY AIR**. Medical evacuation by air is ideal because it is quick and is frequently easier on the patient. There is NO ABSOLUTE CONTRAINDICATIONS TO EVACUATION BY AIR. There may be restrictions placed on the aircraft, however. We will discuss these in another class.

8. **ADMINISTRATIVE MANAGEMENT OF CASUALTY EVACUATION (CASEVAC REPORTING PROCEDURES).** (FMST.07.17b) The system currently used in the FMF for combat conditions and in peacetime training exercises is outlined in the Marine Corps Order P3040. This system reports all personnel losses, regardless of cause, to a central location, utilizing a standard format. There are five steps in this CASEVAC reporting chain, they are as follows:

a. The first link in the chain is the I.D. Tags (Dog Tag):

(1) Everyone must have a chain and two tags.
(2) Those with allergies must have a RED MEDICAL warning tag (NAVMEDCOMINST 6150.29).

(3) Both tags remain with the patient at all times.

b. The second link of the chain is the U.S. Field Medical Card (DD Form 1380):

(1) Must be filled out as completely as possible AFTER the initial first aid is rendered.

(2) Record all treatment and medications and the time given.

(3) The original card remains with the casualty until he reaches his ultimate destination in the chain of evacuation.

c. Casualty Reporting Procedures Manual (MCO P3040.4B w/change 1):

(1) Provides a standard format for reporting all casualties.

(2) A coordinated system MUST be established between medical and S-1.

(3) Type the format on standard bond paper.

(4) Leave enough space to plainly print all information.

(5) Establish a logbook to assign sequential number.

1. Log out at medical, log in at S-1.

(6) Use an asterisk for the medical information portion.

(7) Section 4009 is the reporting format for hostile and non-hostile casualties; sections 4010 and 4011 give additional guidelines.

d. Heat/Cold Casualty Report, (NAVMED 6500/1), if required (NSN 0105-LF-206-5505):

(1) Self explanatory.

(2) Send one copy directly to NAVMEDCOM.

(3) Send the other copies as directed.

e. Local accident report forms if required:

(1) MCO P3040.4B reporting procedures may be sufficient.
(2) Local commands may have separate/additional accident reports that may be required.

**UNITED STATES MARINE CORPS**
Mountain Warfare Training Center
Bridgeport, California 93517-5001

FMST.07.13
05/01/01

## STUDENT HANDOUT

## HIGH-ALTITUDE ILLNESSES

**TERMINAL LEARNING OBJECTIVE:** Given a unit in a cold weather or wilderness high altitude environment , and necessary equipment and supplies, manage high altitude health problems, to prevent death or further injury per the references. (FMST.07.13)

**ENABLING LEARNING OBJECTIVES:**

(1) Without the aid of references, choose from a given list the correct definition of acute mountain sickness, in accordance with the references. (FMST.07.13a)

(2) Without the aid of references, select from a given list the four main mechanisms of acute mountain sickness, in accordance with the references. (FMST.07.13b)

(3) Without the aid of references, from a given list choose the most common and prominent symptom of acute mountain sickness, in accordance with the references. (FMST.07.13c)

(4) Without the aid of references, select from a given list the field management of acute mountain sickness, in accordance with the references. (FMST.07.13d)

(5) Without the aid of references, select from a given list three of the seven high altitude health preventive measures, in accordance with the references. (FMST.07.13e)

(6) Without the aid of references, select from a given list the correct definition of high altitude cerebral edema, in accordance with the references. (FMST.07.13f)

(7) Without the aid of references, select from a given list the pathophysiology of high altitude cerebral edema, in accordance with the references. (FMST.07.13g)

# 21

(8) Without the aid of references, select from a given list three of the symptoms of high altitude cerebral edema, in accordance with the references. (FMST.07.13h)

(9) Without the aid of references, select from a given list the field management of high altitude cerebral edema, in accordance with the references. (FMST.07.13i)

(10)Without the aid of references, select from a given list the correct definition of high altitude pulmonary edema, in accordance with the references. (FMST.07.13j)

(11)Without the aid of references, select from a given list three of the nine symptoms of high altitude pulmonary edema, in accordance with the references. (FMST.07.13k)

(12)Without the aid of references, select from a given list the field management of high altitude pulmonary edema, in accordance with the references. (FMST.07.13l)

(13)Without the aid of references, select from a given list choose the most important treatment for both high altitude cerebral edema and high altitude pulmonary edema, in accordance with the references. (FMST.07.13m)

**OUTLINE**

1. **ACCLIMATIZATION**.

a. Successful acclimatization depends on three primary factors:

(1) Degree of hypoxic stress (i.e. altitude)

(2) Rate of onset of hypoxic stress (i.e. ascent rate)

(3) Individual Physiology (i.e. genetic differences between individuals)

b. Physiologic Changes:

(1) Respiratory - Hypoxic Ventilatory Response (HVR)

- Hypobaric hypoxia
- Triggers carotid body
- Central respiratory center stimulated
- Ventilatory rate increases

HVR can be measured:

- "High" leads to successful acclimatization and superior performance.
- "Low" tends to be found in victims of altitude illnesses.
- HVR can be influenced by depressants (ETOH) and stimulants. (Caffeine)

- May be negatively influenced by physical conditioning.

(2) Respiratory – Fast response: minutes to hours

- HVR leads to increased Ventilatory rate.
- Increased ventilations "blows off $CO_2$."
- Low $CO_2$ levels yield a respiratory alkalosis and rising pH.
- Elevation of pH feeds back to depress the central respiratory center.
- No further immediate rise in ventilatory rate (A brake is placed on any further increase in respiration.)

(3) Renal – Slower response: kidneys respond over the next 24 - 48 hours:

- Kidneys respond to alkalosis by inducing a bicarbonate diuresis.
- PH slowly declines - brake on central respiratory center eases.
- Respiratory rate gradually rises over 4 - 7 days to new baseline.
- Final respiratory rate is altitude dependent.

(4) Cardiovascular - changes are a result of increased sympathetic tone (due to hypoxia)

- Increased: Blood Pressure, venous tone and Cardiac output (increased heart rate and contractility.)
- Stroke Volume declines in first 1 - 3 days due to bicarb diuresis.
- Pulmonary Capillary Wedge Pressure low due to decreased Stroke Volume

(5) Pulmonary Circulation - critical in understanding pathophysiology of HAPE

- Ascent results in global Hypoxic Pulmonary Vasoconstriction.
- Elevated Pulmonary Artery Pressure (PAP) occurs.
- Exertion at altitude worsens this Pulmonary Hypertension.
- Damage to pulmonary vascular endothelium leading to leaky capillaries.

(6) Cerebral Circulation - critical in understanding pathophysiology of HACE

- Hypoxia -----------Vasodilatation
- Hypocapnia ----------Vasoconstriction
- Hypoxia is primary influence resulting in vasodilatation.
- On average, Cerebral Blood Flow (CBF) increases 24% on ascent.
- Extraction Fraction increases at higher altitudes.

(7) Hemopoietic Response

- Erythropoietin levels increase within two hours of ascent

- New immature Red Blood Cells (RBC) seen within days
- New mature RBC's present within 4-5 days
- RBC mass increase seen over weeks to months (altitude dependent)
- Early changes due to hemoconcentration
- Polycythemia is a potential danger

(8) Muscle Tissue

- Intramuscular capillary density increases after three weeks exposure to altitude.

(9) Sleep Pattern

- Stage 1 increase, Stage 2 unaffected
- Stage 3 & 4 decrease
- Rapid Eye Movement (REM) decrease
- Arousals increase: Explanation: Periodic Breathing (Cheyne-Stokes)
- Nocturnal.
- Increased respiratory rate leads to respiratory alkalosis
- Elevated pH depresses central respiratory center
- Apnea
- $O_2$ saturation declines, $CO_2$ levels rise
- Carotid body stimulated
- Ventilatory rate rises

(10)Summary: Physiologic changes seen at altitude result from the body's responses to hypobaric hypoxia. It has been speculated that man would be limited to an 8000-foot elevation limit without supplemental oxygen if not for the acclimatization process. As Messner and Haebler demonstrated on Mt. Everest, acclimatization enables man to explore the entire planet.

2. **<u>ACUTE MOUNTAIN SICKNESS (AMS)</u>**.

a. Definition. AMS is an acute, self-limiting illness, which results when an unacclimatized individual ascends rapidly to high altitude. (FMST.07.13a)

(1) It is rare below 8,000 feet but will occur in many persons rapidly exposed to altitudes greater than 10,000-12,000 feet in elevation.

(2) Aerobic fitness is no predictor for risk of AMS with exposure to altitude. There is some evidence that prior aerobic fitness leads to increased AMS incidence, most likely due to excessively rapid ascent or over-exertion at altitude. Elite athletes may also be at increased risk due to depressed cardiovascular and ventilatory responses to exertion.

(3) Prior ascents to altitude without symptoms of AMS are no guarantee against having symptoms of AMS with future ascents. However, those with past AMS are at an increased risk of AMS with exposure to high altitude. Generally, there is no correlation between the severity of the illness and increasing altitudes. The incidence is slightly increased in younger persons.

(4) There is still much we do not know about altitude-induced physiologic changes. Current explanations in the literature center around genetic differences in hypoxic ventilatory response and vascular endothelium release of inflammatory mediators and Nitric Oxide.

b. Pathophysiology of AMS: The physiologic changes associated with the symptom complex known as AMS are linked to: (FMST.07.13b)

(1) Poor ventilatory response: An individual's Hypoxic Ventilatory Response (HVR) dictates how well he/she will respond to the hypobaric hypoxia of altitude. Those with a brisk HVR will do well, whereas those with a blunted response tend to suffer the symptoms of AMS. A poor ventilatory response leads to hypoxia and hypercapnia (elevated C02 levels). These two combine to augment cerebral blood flow by inducing cerebral vasodilatation.

(2) Fluid retention: The increased sympathetic tone associated with hypobaric hypoxia explains many of the changes seen in the renal system at altitude. Low Renal Blood Flow, increased anti-diuretic hormones, low Glomerular Filtration Rates, and decreased Urine Output all contribute to a Net Fluid Gain. This retained fluid redistributes within the body's fluid spaces. As a result, intracellular and extravascular fluid shifts occur, especially in the brain, the lung, and the interstitial tissues in the periphery.

(3) Increased Intra-cranial Pressure: Increased cerebral blood flow in combination with fluid shifts leads to increasing intra-cranial pressure and associated symptoms. Increased CSF production has also been observed in individuals at altitude.

(4) Altered Pulmonary Mechanics. On ascent to altitude, there is an increase in extravascular lung water. This has been demonstrated via Pulmonary Function Tests showing decreased Vital Capacity, decreased Peak Expiratory Flows, and Increased Alveolar-Arterial Gradients. One study showed a 54% increase in estimated lung mass. It is not uncommon to hear crackles in all lung fields when auscultating asymptomatic individuals. The presence of intra-alveolar fluid inhibits the diffusion of oxygen across the alveolar membrane, thereby reducing oxygenation across the pulmonary capillary membrane.

Signs and Symptoms of AMS. These tend to occur within the first 1- 3 days of exposure to altitude. Symptoms should resolve spontaneously within 15 - 24 hours, if ascent is arrested. The three most prominent symptoms are Headache, Nausea and Vomiting. (FMST.07.13c) Although nonspecific, headache must be present to make a diagnosis of AMS. 95% of AMS victims will complain of one or more of these three. Other common complaints include dizziness, anorexia, drowsiness, malaise, weakness, and insomnia. It is possible that findings of dyspnea on exertion and mental status changes along with ataxic gait may accompany the classic AMS signs and symptoms. These are clear indications for descent, as these are the hallmark signs of progression to HACE and the development of HAPE. Again, headache is the most common and prominent symptom of AMS.

c. Field Management (FMST.07.13d)

(1) Stop any further ascent. Light duty. No tobacco.

(2) Symptomatic Treatment. Mild analgesics such as Aspirin or Tylenol for the headache, Compazine or Phenergan for nausea/vomiting.

(3) Other Meds: Acetazolamide (Diamox) Dexamethasone (Decadron) Diuretics (last choice) Oxygen.

(4) Hyperventilate: It is unrealistic to try to consciously control one's respiratory rate over any appreciable period of time. It is nearly impossible to maintain conscious hyperventilation while functioning in the field.

(5) Descend. If none of these methods help, then individuals should descend 1,000-3,000 feet. This will usually result in marked relief of symptoms. Most cases, however, improve in 1-2 days with the above measures.

(6) ALL PATIENTS WITH AMS MUST BE EVALUATED FOR HAPE AND HACE.

HOW?

- Field sobriety test.
- Mini-mental status exam.
- Auscultation of lung fields.

d. Preventive Measures: (FMST.07.13e)

(1) The best approach to high-altitude travel is staged, gradual ascent, when moving at altitudes greater than 3000 meters. Gradual ascent means no faster than 3000 ft/day up to 14,000 and no greater than 1,000 ft/day over 14,000 ft.

(2) Avoid any alcohol and/or sedatives during the first two nights.

(3) Maintain adequate fluid intake.

(4) Work high - Sleep low.

(5) High Carbohydrate Diet (approximately 70% of diet).

(6) Avoid overexertion, but remember that mild exercise assists acclimatization.

(7) No smoking.

(8) Pharmacological options: Prevention of AMS is possible with Acetazolamide (Diamox). It dramatically lowers the incidence of AMS when taken prophylactically. The usual dose is 125 mg PO BID starting 24-48 hours before ascent and should be continued for 3-4 days after the start of the mission. It is also effective as a treatment, after symptoms have already begun.

3. **HIGH ALTITUDE CEREBRAL EDEMA (HACE)**.

a. Definition. HACE is a high-altitude illness, which is characterized by swelling of the brain. (FMST.07.13f)

(1) HACE can occur as low as 8,000 feet, but typically occurs at more than 12,000 feet.

(2) HACE is considered a progression of AMS, usually as a result of proper precautions not being adhered to with regard to ascent rate, sleep altitude, over-exertion and especially, continued ascent despite signs and symptoms of AMS.

b. Pathophysiology of HACE. HACE is considered a progression of AMS. The increased cerebral blood flow along with the fluid shifts mentioned above lead to vasogenic edema (leaky blood vessels) inside the confined space of the skull. This interstitial edema increases the diffusion distance of oxygen between the capillaries and brain cells, leading to Ischemia (poor oxygen delivery), cellular swelling results. Once this pressure nears arterial pressure, cerebral blood flow will be impaired. Eventually, intra-cranial pressure will rise with a concomitant increase in the risk of brain herniation leading to, cranial nerve palsies, or strokes. Increased CSF production further compounds the problem. (FMST.07.13g)

c. Differential Diagnosis: Other illnesses are possible in individuals ascending to altitude. Your differential should include the following when neurologic deficits are involved:

(1) Cerebrovascular spasm and Transient Ischemic Attacks.

(2) Cerebral vascular thromboembolic events.

(3) Intra-cranial hemorrhage (aneurysm, AVM).

(3) Hypothermia.

(4) Meningitis

(5) Encephalitis

(6) "Hangover"

d. Signs and Symptoms:

(1) Early signs and symptoms are those found in AMS. (i.e. H/A, NV)

(2) When the above warning signs go unheeded the true signs and symptoms of HACE can quickly follow: (FMST.07.13h)

(a) Ataxia (loss of muscle coordination leading to difficulty maintaining balance), especially prominent in the victim's gait.

(b) Mental Status Changes, poor judgement, personality change which may progress to stupor, coma and death relatively rapidly without proper treatment

(c) Lassitude, confusion, hallucinations, convulsions, behavioral aberrations.

(d) Paralysis of one or more extremities (see Differential Diagnosis).

(3) The most important impediment to early recognition of HACE is its insidious onset. Early signs and symptoms frequently go unrecognized or are ignored by patients, as well as their companions, who also may be suffering to some degree from the effects of altitude.

e. Field Treatment (FMST.07.13i)

(1) Early recognition is the KEY. Look for Ataxia and Mental Status Changes.

(2) Treatment should be immediate since fatalities can occur within a few hours in severe cases.

(3) Once diagnosed, the patient should be placed in the most comfortable position possible, descent should be directed immediately and administration of high concentration O2 should be given, in route, if available.

(4) Decadron (early), Diamox, 10mg IM/IV and/or Diuretics are all acceptable medical interventions. (Be careful with diuretics!)

(5) If loss of consciousness, think airway management and ABC's.

(6) Casevac to an advanced medical facility ASAP!

(7) A Gamow Bag can be a lifesaver if decent is not immediately possible.

**NOTE:** The Gamow Bag is a one-man portable hyperbaric chamber. It weighs 14.5 lbs., is operated by a foot-pump, and is made of nylon with an airtight zipper. With the HAPE or HACE victim inside, the bag is pressurized with the foot-pump to an internal pressure of 2 psi. This is the equivalent of a 5000 - 6000 foot descent in regards to the partial pressure of oxygen inside the bag. Air is circulated within the bag via the foot-pump. This can be a life-saving resource in the event a victim is caught at altitude and cannot descend due to weather or other uncontrollable variables.

f. Prevention

(1) Preventive measures are the same as those discussed for AMS.

(2) Decadron can be used as a prophylactic medication along with Diamox, in cases of unavoidable rapid ascent to extreme altitude. However one must continue this regimen, while at altitude, or face rapid onset of symptoms once the medication is stopped.

4. **HIGH ALTITUDE PULMONARY EDEMA (HAPE).**

a. Definition. HAPE is a high-altitude illness that is characterized by filling of the lungs with edema fluid. (FMST.07.13j)

(1) HAPE rarely occurs below 12,000 feet.

(2) HAPE victims in the past were commonly described, after autopsy, as victims of "pneumonia" and "Inflammation of the lungs". Since 1960, medical science has had the technology to investigate the complexities of the physiological changes associated with hypobaric hypoxia.

(3) Incidence:

1:10,000 Colorado skiers
1:50 Mt. McKinley climbers
1:7 Indian soldiers (Indo-China Border Conflict)

(4) Commonly occurs within 2-4 days of ascent.

(5) Risk for developing HAPE remain same as above:

(a) Rate of ascent
(b) Sleeping altitude
(c) Level of exertion
(d) Gender (M>F)
(e) Age (young>old)
(f) Individual physiology

b. Pathophysiology of HAPE:

(1) Hypoxic Pulmonary Vasoconstriction

(a) Normal part of the acclimatization process (elevated PAP)
(b) Adaptive at sea level (i.e. lobar pneumonia)
(c) Maladaptive at altitude (global hypoxia)
(d) Tends to be uneven, leading to high pressure/high flow area.

(2) Overperfusion Theory

(a) Localized areas exposed to high pressure and high flow
(b) Shearing forces and stress failure at the microvascular level
(c) Endothelial damage to capillaries permits high molecular weight protein leak.
(d) Combines with hydrostatic pressure leakage of intravascular water.

(3) Nitric Oxide deficiency theory

(a) HAPE prone individuals may have deficiency of NO, a potent vasodilator, synthesized within pulmonary vascular endothelium.
(b) Deficiency of NO in the pulmonary vascular bed results in global Hypixic pulmonary vasocostriction.
© High pulmonary artery pressures result in hypertensive mediated damage to vascular endothelium with release of inflammatory mediators.
(c) Inflammatory mediators lead to increased vascular permeability.
(d) Net result is pulmonary edema from increased vascular permeability and high pulmonary hydrostatic pressures.

c. Pathologic Findings of HAPE: Postmortem studies of victims of HAPE have yielded some interesting results:

(1) Grossly, the lungs are congested and swollen, as seen with other causes of pulmonary edema. Average lung weight on autopsy is 2-4 times normal. Histologically, one sees pulmonary edema with a protein rich exudate filling the alveoli. It is also common to find the alveolar spaces filled with hyaline membranes.

(2) The left heart is completely normal in appearance. However, the right ventricle, right atrium, and pulmonary arteries are distended and dilated.

(3) Greater than 50% of HAPE victims are found to have evidence of HACE.

d. Signs and Symptoms. These tend to occur within 2-4 days of arrival at altitude. Usually the symptoms of AMS are present before or occur with the symptoms of HAPE. (FMST.07.13k)

(1) Early signs:

(a) Dry cough, frequently occurring at night.

(b) Dyspnea on Exertion (DOE), especially with ambulation uphill.

(c) Mild chest pain - usually perceived as an ache beneath the sternum.

(d) Decreased work performance and increased recovery time between events.

(e) Peripheral or Central Cyanosis.

(2) Later signs:

(a) Dyspnea at Rest.

(b) Productive cough which yields large amounts of pink, frothy sputum.

(c) Rapid pulse and respiratory rates.

(d) Audible crackles on auscultation.

(e) Mental status changes, Ataxia, Loss of Consciousness.

(f) CXR findings:

<u>ECHO findings:</u>

-Non-cardiogenic pulmonary edema -increased PAP

-Tricuspid regurgitation.
-Normal left ventricle
-Prominent pulmonary arteries -enlarged right atrium
-Enlarged right ventricle.

**NOTE:** Crackles heard on lung exam, as an isolated finding is NOT an indication for descent or grounds for the diagnosis of HAPE. It is common (up to 15% of climbers on Mt. Rainier) to find asymptomatic rales in climbers at significant altitude. Concern is heightened when crackles occur in constellation with other signs and symptoms.

e. Field Treatment (FMST.07.131)

(1) The most important emergency care measure is immediate descent to a lower altitude, since fatalities can occur within 6-12 hours in severe cases.

(a) A descent of at least 2,000-3,000 feet can be a definitive treatment if the condition is caught early enough.

(2) The patient should be placed in the most comfortable position (usually sitting) and given high-flow O2 if available. It is important to minimize activity, as exertion increases pulmonary artery pressures.

(3) Medication regimens include:

(a) Nifedipine (10mg PO initially then 30mg SR Q12hr) - pulmonary artery/arteriole dilator.

(b) Diamox and Decadron may be appropriate regimens, depending on your medication availability.

(4) Oxygen is always appropriate, however, it is not always available.

(5) Gamow Bag can be a lifesaver if decent is not immediately possible.

f. Prevention Measures. Preventive measures are the same as those discussed for AMS.

(1) Acetazolamide (Diamox) has been proven to be successful in the prevention of HAPE. 125 - 250 mg po bid helps prevent HAPE in individuals with recurrent episodes.

(2) It should be noted that a history of HAPE is NOT a contraindication to the individual participating in high-altitude activities. Each case should be evaluated individually.

**NOTE:** Remember decent is the most important treatment for both HACE and HAPE. (FMST.07.13m)

5. **SYSTEMIC EDEMA AT HIGH ALTITUDES**.

Occasionally, certain individuals, especially women, will develop edema of the face, hands, and feet at altitude. This edema usually tends to occur in the absence of other symptoms. In susceptible persons, repeat episodes are common. The edema goes away upon return to lower altitude. There is no need for descent in these cases because the patient is usually able to tolerate the edema and continue to function well. However, they should be examined carefully to rule out the presence of HAPE or HACE. Diuretics are usually not warrented. It is also probably reasonable to restrict salt intake. The only indication for descent is if the edema is causing functional disability, i.e. inability to don one's boots, or facial edema severe enough to restrict vision.

6. **SICKLE CRISIS**.

Persons with sickle cell trait are at risk for developing sickling attacks when exposed to hypoxic stress, as might occur at high altitudes. However, sickling can occur at lower elevations. Individuals with Mediterranean family ties should be considered at risk for having sickle cell trait. The trait is more likely in black individuals as well. Any patient complaining of Left Upper Quadrant pain, SOB, and/or arthralgia at altitude should have sickling crisis included in the differential diagnosis regardless of race. It has been known to occur in white females as well. It has been recommended by some that individuals with sickle cell trait should be restricted to an altitude of 8,000 feet and below. However, more liberal recommendations exist, suggesting that ultimate altitude be dictated by the individual and their symptoms.

7. **ALTITUDE THROAT**.

At higher altitudes, mouth breathing is often increased during exertion. Inhalation of this cold, dry air can result in drying of the mucous membranes in the throat and upper airway. This condition responds to humidified air, salt-water gargles or lozenges. If exudate or edema is present, throat culture and/or antibiotics may be warranted.

8. **HIGH ALTITUDE FLATUS EXPULSION HAFE)**.

Ascent to altitude can result in rapid expansion of intestinal gas, due to the decreased pressure of altitude. This produces abdominal discomfort and the passage of colonic gas. This condition is not associated with serious consequences, however, it may create tension amongst tent-team members. Treatment consists of oral administration of anti-flatulence such as simethicone, Mylanta II, or Gaviscon. Know and avoid your gas producing foods!

9. **HIGH ALTITUDE RETINAL HEMORRHAGE**.

Up to 60% of individuals ascending to altitudes greater than 18,000 feet will develop asymptomatic retinal hemorrhages. On fundoscopic exam, there will be hyperemia of the disc and dilation and tortuosity of the vessels. "Flame" hemorrhages and "cotton wool spots" can be seen on exam. The only indication for descent in an individual with retinal hemorrhages is loss of visual acuity. This usually results when a large hemorrhage involves the macular region of the retina. Remember that Carbon Monoxide poisoning can present with retinal hemorrhage as well.

10. **THROMBOEMBOLIC EVENTS**.

Thromboembolic disease occurs with increased frequency at higher altitudes. It is believed that the common circumstances of volume depletion due to dehydration and Polycythemia due to altitude exposure predispose individuals to coagulopathic events. In addition, extended periods of inactivity due to "going to ground" during storms contributes to this predisposition. There does not seem to be any direct influence of hypobaric hypoxia on the body's normal coagulation system.

11. **ACETAZOLAMIDE**.

Acetazolamide (Diamox) is a Carbonic Anhydrate Inhibitor, which is especially active in the brain, the lung, and the kidney.

a. Kidney:

(1) Uptake of bicarbonate ion is inhibited.

(2) Bicarbonate diuresis is induced.

(3) Metabolic acidosis ensues within one hour of medication.

(4) Sequence mimics and accelerates the natural acclimatization process.

(5) Enhanced HVR is the end result.

(6) Diuresis addresses problems associated with fluid retention.

b. Brain:

(1) Direct effects at the level of the Central Respiratory Center.

(2) Diminishes periodic breathing during sleep.

(3) Fewer episodes of apnea and extreme hypoxemia.

(4) CSF production is inhibited and CSF absorption is enhanced.

c. Side Effects:

(1) Polyuria.

(2) Flattens taste of carbonated beverages.

(3) Peripheral Paraesthesias, myopia, impotence, nausea, and drowsiness.

(4) Contraindicated for Sulja Allergic and G6PD-deficient individuals (reversible when MED. s discontinued.)

**UNITED STATES MARINE CORPS**
Mountain Warfare Training Center
Bridgeport, California 93517-5001

FMST.07.02
08/21/01

# STUDENT HANDOUT

# ENVIRONMENTAL HAZARDS (AVALANCHE)

**TERMINAL LEARNING OBJECTIVE**. Given a unit in a cold weather environment and necessary equipment and supplies, assess winter environmental hazards, in accordance with the references. (FMST.07.02)

**ENABLING LEARNING OBJECTIVES**.

(1) Without the aid of references, select from a given list the different types of avalanches, in accordance with the references. (FMST.07.02a)

(2) Without the aid of references, select from a given list the description of Equilibrium Growth Form (ET), in accordance with the references. (FMST.07.02b)

(3) Without the aid of references, select from a given list the relative stability of Equilibrium Growth Form (ET), in accordance with the references. (FMST.07.02c)

(4) Without the aid of references, select from a given list the description of Kinetic Growth Form (TG), in accordance with the references. (FMST.07.02d)

(5) Without the aid of references, select from a given list the relative stability of Kinetic Growth Form (TG), in accordance with the references. (FMST.07.02e)

(6) Without the aid of references, select from a given list the Melt-Freeze Metamorphism, in accordance with the references. (FMST.07.02f)

(7) Without the aid of references, select from a given list the relative stability of Melt-Freeze Metamorphism, in accordance with the references. (FMST.07.02g)

(8) Without the aid of references, select from a given list how wind effects the slopes in snow covered mountainous terrain, in accordance with the references. (FMST.07.02h)

**22**

(9) Without the aid of references, select from a given list the route considerations, in accordance with the references. (FMST.07.02i)

(10)Without the aid of references, select from a given list the individual preparations to take when crossing a potential avalanche prone slope, in accordance with the references. (FMST.07.02j)

## OUTLINE

1. **DEFINITION**. Avalanches are falling masses of snow that can contain rock, soil or ice, which will travel over terrain of least resistance.

2. **AVALANCHE PATH**. The term avalanche path defines the area in which an avalanche runs. An avalanche is generally divided into three parts: Starting Zone, Track, and Runout Zone.

(1) Starting Zone. This is where the unstable snow fell and began to move.

(2) Track. This is the slope or channel down, which snow moves at a fairly uniform speed.

(3) Runout Zone. This is where the snow slows, debris is deposited, and the avalanche stops.

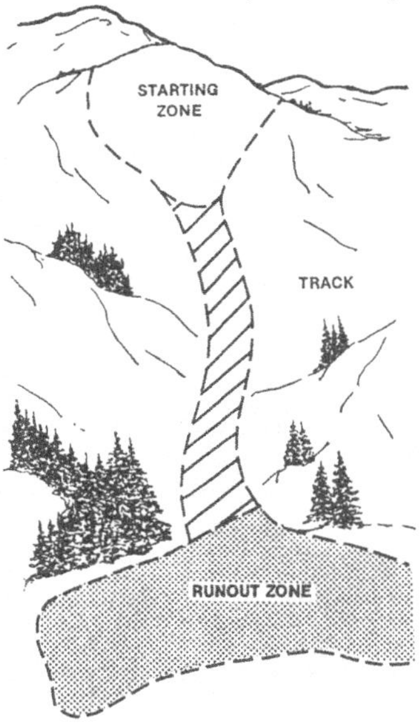


AVALANCHE ZONES

3. **TYPES OF AVALANCHES**. (FMST.07.02a)

a. <u>Loose Snow Avalanches</u>. Loose snow slides, also called point releases, start with a small amount of cohesionless snow and typically pick up more snow as they descend. From a distance, they appear to start at a point and fan out into a triangle. They usually are small, involving only upper layers of snow, but they are capable of being quite large and destructive depending upon how much material they entrain. Factors of a loose snow avalanche are:

(1) The stress of the moving snow in a loose snow slide can also trigger larger and deeper slab releases.

(2) Loose snow releases occur most often on steep slopes of 35° or higher.

(a) During or shortly after a snowstorm.

(b) During warming events caused by rain, rising temperatures, or solar radiation.

LOOSE SNOW AVALANCHE

b. <u>Slab Avalanche</u>. Slab avalanches occur when one or more layers of cohesive snow break away as a unit. As the slab travels down slope, it splits up into smaller blocks or clods. Factors of a slab avalanche are:

(1) Slab failure is commonly initiated when the bond between the slab and the bed surface fails, thus placing tremendous stress on the other boundary regions, which, in turn, are unable to hold the slab in place.

(2) Slab thickness can range from less than an inch to 35 feet or more, and range in width from a few yards to well over a mile.

(3) Slab material is also highly variable. Slabs may be hard or soft, wet or dry.

(4) The speed of a slab avalanche can range from roughly 65 mph for a wet slab, on up to 150 mph for a dry slide.

(5) Most slab avalanches release on slopes with angles between 35° and 40°.

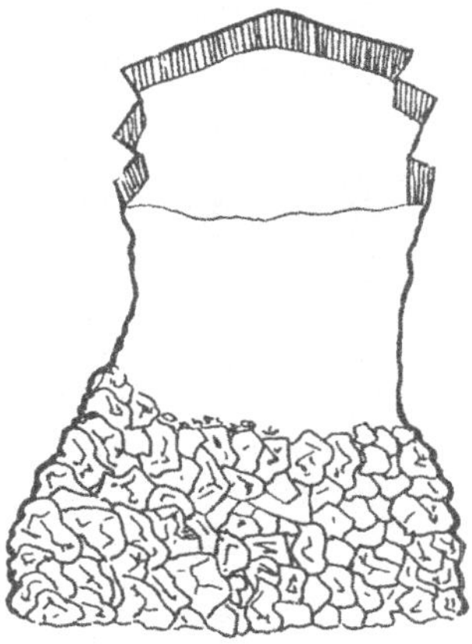

SLAB AVALANCHE

(6) Nomenclature of a Slab Avalanche

(a) Crown. This is the breakaway wall of the top periphery of the slab. It is usually at a right angle to the bed surface. It is formed by tension fracture through the depth of the slab from bottom to top.

(b) Bed Surface. This is the surface over which the slab slides. The bed surface can be the ground.

(c) Flanks. These are the left and right sides of the slab.

(d) Stauchwall. This is the lowest downslope fracture surface. The slab material usually overrides it and it consists of a diagonal shear fracture of wedge-like shape.

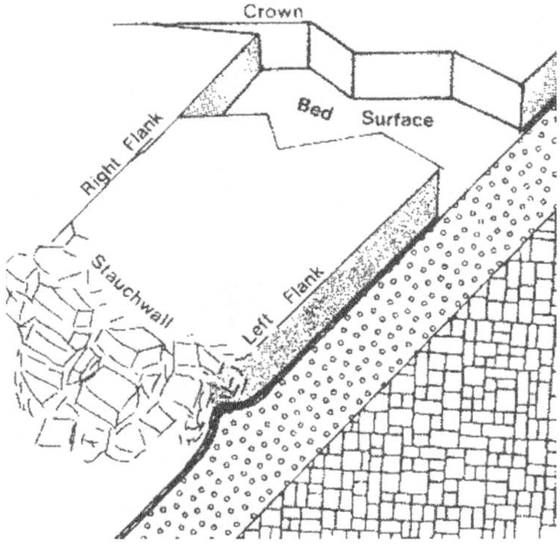


NOMENCLATURE OF SLAB AVALANCHE

c. Cornice Collapses. Cornices form when windblown snow builds out horizontally at sharp terrain breaks such as ridgecrests and the sides of gullies. These cornices can break off well back from their edges. They often trigger bigger slides when they hit the wind-loaded, pillowed area on the slope below.

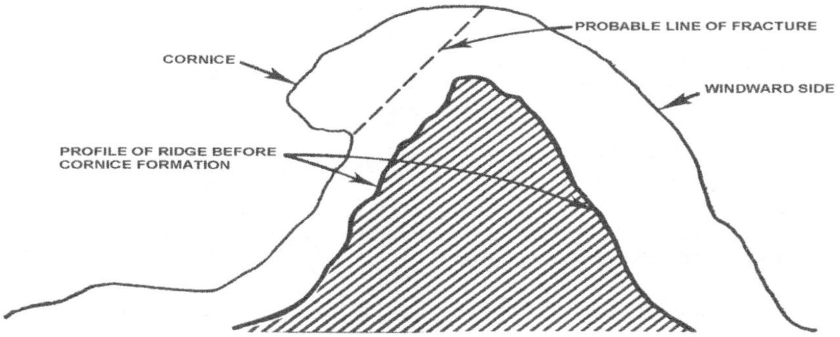


CORNICE

d. Ice Avalanches. Ice avalanches are caused by the collapse of unstable ice blocks (seracs) from steep or overhanging part of a glacier. Ice avalanches can entrain a considerable amount of rock, ice, and snow and travel long distances.

4. **AVALANCHE HAZARD EVALUATION PROCESS**.

a. The Data Triad. The interaction of three critical variables will help in determining whether or not an avalanche is possible. These three variables are:

(1) The Terrain.

(a) Terrain Analysis. Is the terrain capable of producing avalanches?

(2) The Snow pack.

(a) Stability Evaluation. Could the snow slide?

(3) The Weather.

(a) Avalanche Forecasting. Is the weather contributing to instability?

b. To determine whether an avalanche hazard exists, we must add another variable; humans. Without the presence of humans, there is no hazard.

(1) The Human Factor.

(a) Decision Making. What are you alternatives and their possible consequences?

5. **TERRAIN**. Learning to recognize avalanche terrain is the critical starting point in the avalanche hazard evaluation process. Assuming that avalanches occur on only big slopes is a very common mistake. The following factors influence whether a given slope is capable of producing an avalanche and will help you recognize avalanche terrain:

a. Slope Angle: Slope angle is the most important terrain variable on determining whether or not it is possible for a given slope to avalanche. The underlying concept is that as the slope angle increases, so does the stress exerted on all boundary regions of the slab.

(1) Slope angles less then 25° will rarely slide due to lack of stress to the snowpack.

(2) Slab avalanches in cold snow are possible between the slope angles of 25° and 60°.

(3) Most slab avalanches release on slopes with starting zone angles between 35° and 40°.

(4) Slope angles 60° or greater will continually sluff due to large amounts of stress to the snowpack.

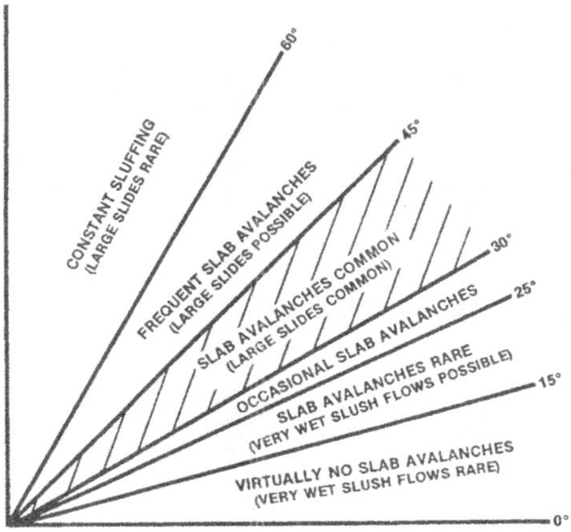


SLOPE ANGLE RELATIONSHIP TO AVALANCHES

**NOTE**: It is important to know that your motion or body weight can trigger an avalanche even if you are on a low angle slope or on the flats as long as this terrain is connected to a slope with a angle of roughly 25° and instability exists.

b. Slope Aspect (Orientation): Subtle changes in slope aspect can greatly affect snow stability.

(1) Leeward. Deposition of wind-transported snow increases the stress on the snow pack and enhances slab formation.

(2) Moderate warming by the sun can help strengthen and stabilize the snowpack.

(3) Intense, direct sunlight has the opposite effect by weakening and lubricating the bonds between grains.

(4) On shaded slopes, weak layers often persist or are more well-developed because of generally colder conditions and the absence of solar warming during much of the winter. Therefore, suspect instability on shadowed slopes.

c. Terrain Roughness (Anchoring). Slopes with anchors are less likely to avalanche than open slopes.

(1) Boulders, trees, ledges will act as anchors and help hold the snow in place until they are buried.

(2) Smooth slopes (i.e. smooth granite, grassy) may only need 1 foot of snow to release.

(3) Anchors are commonly areas of stress concentration because the snow upslope of them is being held in place while the snow below or to the sides is being pulled downhill by gravity. For this reason, anchors can be starting points for initial failure to occur and fractures often run from tree to tree to rock.

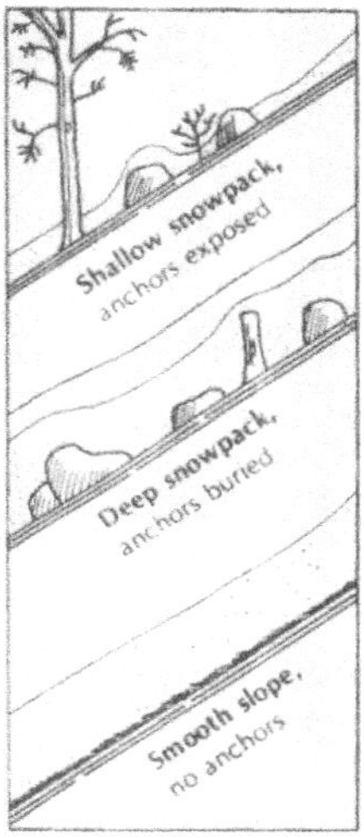


TERRAIN ROUGHNESS

d. Slope Shape. Avalanches can happen on any snow-covered slope steep enough to slide.

(1) Convex Slopes. Slabs are most likely to fracture just below the bulge where stresses are greatest.

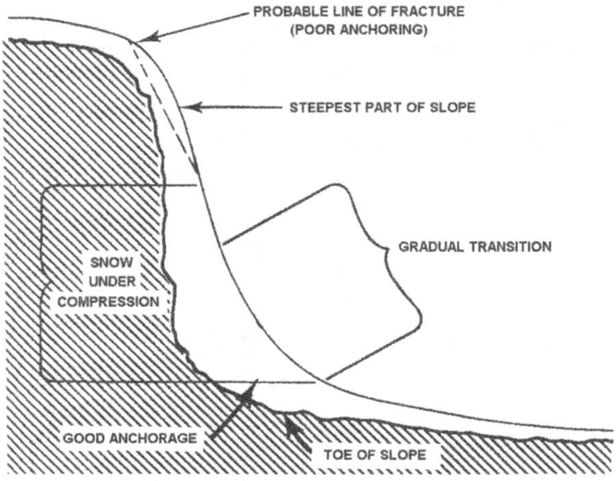


CONVEX SLOPE

(2) Planar Slopes. On these broad, smooth slopes, avalanches can happen anywhere. Slabs often fracture below cliff bands.

(3) Concave Slopes. These slopes provide a certain amount of support through compression at the base of the hollow, but they are still capable of avalanching, especially on large slopes.

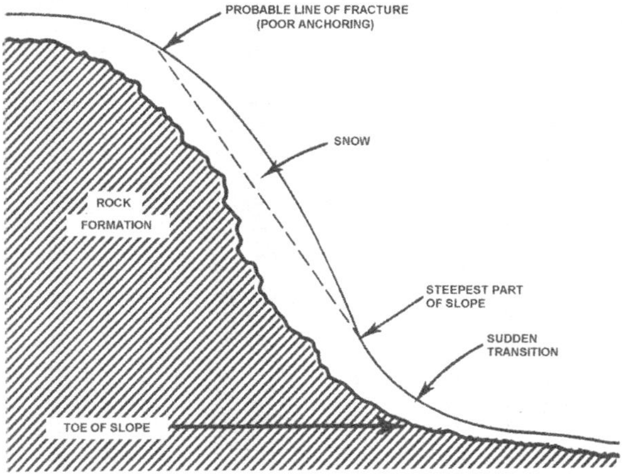


CONCAVE SLOPE

e. Vegetation. Vegetation can provide evidence of both the frequency and magnitude of past avalanche occurrences and thus indicate potential avalanche terrain as well as the capability of a given path. Vegetative indicators include:

(1) Swaths of open slope between forested or vegetated areas.

(2) Trees which are bent, broken, or uprooted, "broomed" trees (i.e., previously broken but with new growth tops), and vegetation, which is polished, or "flagged" (i.e., missing branches on the uphill side). Flagging can also indicate the flow height of the avalanches, which have impacted the area.

(3) Presence of "disaster species" such as alders, willows, dwarf birch and cottonwoods.

(4) Marked difference in height of trees (i.e., smaller spruce in the path, larger on the edges).

VEGETATION INDICATOR OF AN AVALANCHE PRONE SLOPE

f. Elevation. Temperature, wind and precipitation often vary significantly with elevation. Common differences are rain at lower elevations with snow at higher elevations or differences in precipitation amounts or wind speed with elevation. Never assume that conditions on a slope at a particular elevation reflect those of a slope at a different elevation.

g. Path History. All avalanche paths have some sort of history, whether it is their magnitude or to how often they slide. Before going into avalanche country, try and get as much information as possible.

6. **SNOWPACK**. The snowpack accumulates layer by layer with each new snow or wind event. These layers are then subject to changes in texture and strength throughout the winter. The changes help determine snow strength by influencing how well individual snow grains are bonded to each other both within the layer and between layers. Many combinations of strong and weak layers can exist within the snowpack. The structure of the snowpack varies greatly depending upon the particular season, location, climate, slope aspect, inclination and shape.

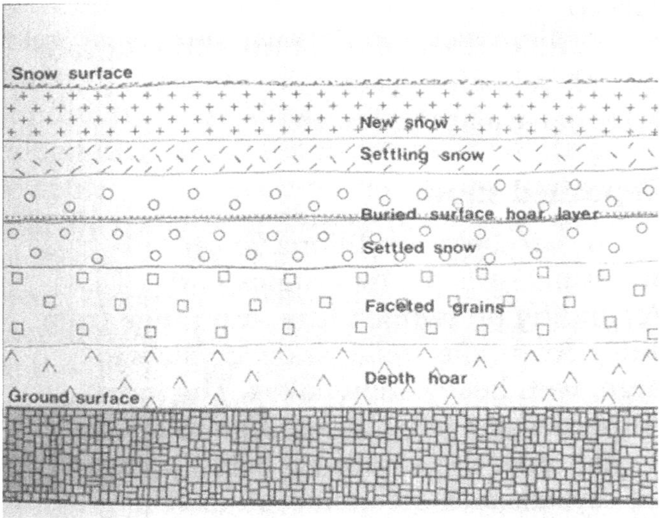


EXAMPLE SNOWPACK

a. Snow Metamorphism. This is the name given to describe the changes in structure that take place over time within the layers of the snowpack. There are several types of snow metamorphism. Each occurs under a different set of conditions and each affects the strength of the snowpack. As conditions change, the dominant type of metamorphism in a given layer may change. Also importantly, different types of metamorphism may be occurring in various layers of the snowpack at the same time. The following are snow metamorphic types encountered in a snowpack:

(1) Equilibrium Growth Form (ET) or Rounding. (FMST.07.02b) Rounded grains develop when temperatures in a layer or between layers are fairly uniform, that is, there is no significant temperature change within the snowpack.

(a) Individual grains become smaller and rounder. Bonds or necks between grains are developed. Thus, the equilibrium form process produces fine, rounded, well-bonded grains and the result is that relatively strong layers form within a snowpack. (FMST.07.02c)

(b) As the equilibrium form process advances in stages, the smaller, rounder, and better bonded the grains will be.

(c) Favorable conditions or habitat for the development of rounded grains are cloudy, mild weather or a thick snowpack.

(2) <u>Kinetic Growth Form (TG) or Faceted Grains</u>. (FMST.07.02d) Faceted grains develop when a significant temperature change exists within or between layers. In most areas, the temperature at the ground/snow interface is warmer than the air temperature. The larger the temperature differences, the quicker the vapor transfer and the process of change.

(a) The shallower the snowpack, the greater the temperature changes within the snowpack. Deep snowpacks tend to dampen this difference by adding many layers of installation between the relatively warm ground and cold air.

(b) The trend of the kinetic growth form process is to produce large, angular grains, which are poorly bonded and weak, especially in shear. This result is that weak layers may form within a snowpack. (FMST.07.02e)

(c) Because faceted grains have as much the same consistency as sugar, they are sometimes referred to as "sugar snow". Advanced faceted grains are also known as depth hoar.

(d) Faceted snow is often the layer that collapses and goes "whump". When subjected to significant loading or wind-transported snow it becomes very sensitive.

(e) Favorable conditions for the development of faceted grains are cold weather and/or a thin snowpack.

(3) <u>Melt-freeze Metamorphism (MF)</u>. (FMST.07.02f) This type of metamorphism occurs during mid-winter thaws or in the spring, when meltwater or rain enters the snowpack and the snowpack temperature reaches 32°F.

(a) The trend is toward the production of coarse, rounded grains and with repeated cycles of melting and freezing. These large grains are also known as "corn snow".

(b) In the freeze phase, these grains are well-bonded and strong creating a stable snowpack. But resulting ice crust can make a good potential bed surface for slabs formed on top of them.

(c) In the melt phase, wet grains weaken rapidly and are lubricated by the presence of free water. This creates an unstable snowpack and is why timing during movement is critical in the spring near steep slopes that are being subjected to warming. (FMST.07.02g)

b. Other Weak Layers. There are some other important weak layers; surface hoar and unmetamorphosed new snow.

(1) Surface hoar is the wintertime equivalent of summertime dew and is formed at the snow surface during cold, clear weather. Surface hoar crystals are loose, feathery, and poorly bonded.

(a) Surface hoar is a potentially deadly weak layer once buried because it persists for a long period of time, can form a thin shear plane that is difficult to detect, and can produce long-running "zipper" fractures.

(2) Unmetamorphosed new snow is snow that may have fallen during cool or windless period of a storm and then had denser, heavier snow deposited on top of it.

(a) These rounded, iced pellets often roll downslope and collect in depressions or at the bottom of cliff bands, thus forming an area that may be more sensitive once the next load is deposited.

7. **WEATHER**. Weather has been termed the "Architect of Avalanches". Most natural avalanches occur during or shortly after storms because the snowpack often cannot adjust to the vast amounts of new weight added in a short time. Weather affects the stability of the snowpack by altering the critical balance between strength and stress. The three main contributing factors are the precipitation, wind, and temperature.

a. Precipitation. The significance of precipitation is that it increases the stress exerted upon a snowpack by adding weight.

(1) Snow. New snow can provide a certain amount of strength to a snowpack, but can also cause rapid loading during storm.

(2) Rain. Heavy rain weakens the snowpack by warming and eroding the bond between grains and slab layers.

b. Wind. Since wind speed and direction help determine which slopes are being loaded this is a very important avalanche consideration. (FMST.07.02h)

(1) Top loading. Wind accelerates on the windward side of terrain features and picks up loose snow, carries it over the crest and deposits it on the leeward side where the wind decelerates and deposits the snow.

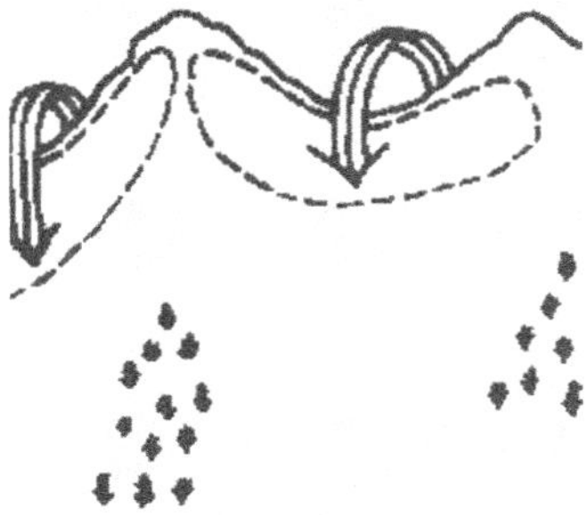

TOP LOADING

(2) Side loading. Also known as cross-loading, is sometimes more insidious because it can be harder to detect, especially in areas of gentle gullies.

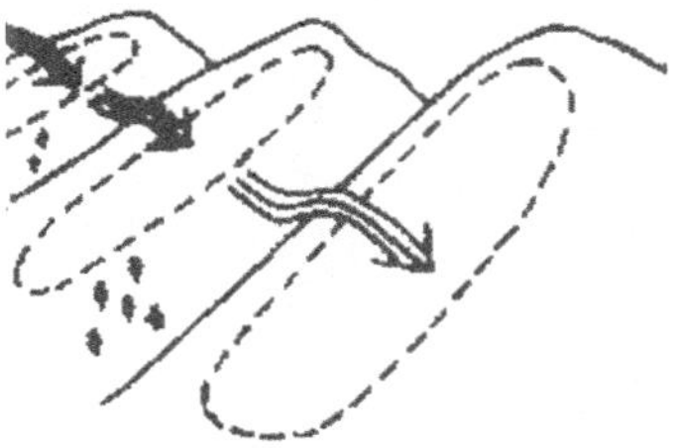

SIDE LOADING

c. Temperature. Changes in snow temperature can significantly affect snow stability. Largely ground and air temperatures, solar radiation and terrestrial radiation govern these changes.

(1) A warm snow pack will settle rapidly, becoming denser and stronger. This is associated with cloudy skies, since clouds trap warm air against the earth's surface.

(2) Though gradual warming encourages strength and stabilization, intense warming weakens the bonds between grains and increasing the rate of the downslope deformation in affected layers.

(3) On shaded slopes, the snowpack not only undergoes less (or slower) settlement due to warming but also the development of weak layers such as faceted snow or surface

hoar. This is because temperature gradients within the snowpack and at the snow surface can be more pronounced and persists for longer periods of time.

(4) Storms that start out cold and get progressively warmer are more likely to produce avalanches than those that start out warm and progressively become cooler.

(5) Any rapid, prolonged rise in temperature following long periods of cold weather could potentially lead to instability and should be noted as one of nature's signs.

8. **HUMAN FACTOR**. It is possible to travel at times of high snow instability by choosing safe routes. Similarly, it is possible to get caught in an avalanche during periods of relatively low snow instability through poor route selection and stability evaluation. In other words, we create potential hazard by travelling in avalanche terrain. But through careful route selection, preparation and decision-making, we can limit the amount of danger that is involved.

a. Route Considerations. Careful route selection can greatly reduce the chances of getting caught in an avalanche and in some areas, make it possible to travel during periods of high instability. The following route considerations should be taken before crossing a potential avalanche prone slope: (FMST.07.02i)

(1) Determine starting zones of probable avalanche prone slopes and cross as high as possible, preferably above natural anchors.

(2) Travel on high points and ridges, especially the windward sides.

(3) When ascending or descending an avalanche prone slope, stay to the sides of the start zone and track.

(4) Avoid wind-loaded, leeward slopes.

(5) Favor terrain with anchors, i.e. tree-covered area, instead of open slopes.

(6) Pick areas with flat, open run-outs so that debris burial depth is decreased. Avoid areas that feed into gullies, crevasses and over cliffs.

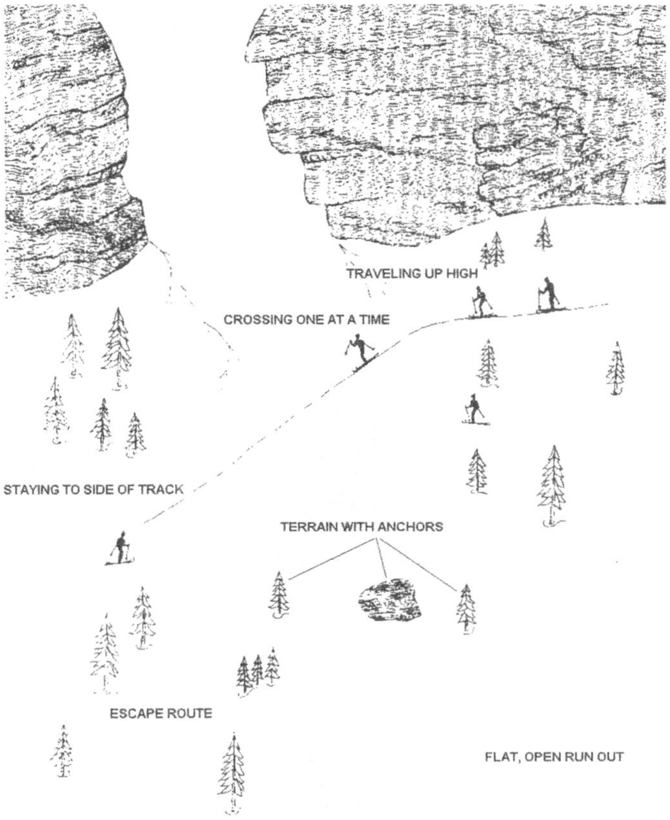


CROSSING AN AVALANCHE PRONE SLOPE

(7) V-shaped and U-shaped Valleys. You can generally find a safe route somewhere in a wide U-shaped valley, but narrow V-shaped valleys should be avoided. In V-shaped valleys, avalanches could run from either side and continue up the opposite side, so there may be little or no safe ground.

b. Crossing Potential Avalanche Slopes. Occasionally, being reasonably sure that you will not fall, or that you can travel quickly, or knowing that all members of the patrol have rescue equipment and are proficient with it may allow you to travel across marginal areas. Before crossing a potential avalanche slope the following preparations should be taken: (FMST.07.02j)

(1) Loosen ski bindings and remove hands from ski pole straps.

(2) Loosen your pack and undo the waist strap.

(3) Secure ECWCS hood tightly covering face and trail a 15 foot avalanche cords if available.

(4) Go straight downhill on foot rather than ski and look for possible escape routes.

(5) Go straight down. Do not traverse.

(6) When possible, cross as high as possible on concave slopes.

(7) Cross one at a time. Just because one crosses it safely doesn't mean that it is safe passage for the rest. Belay everyone else across if possible.

CROSSING AN AVALANCHE PRONE SLOPE

c. Actions if Caught in an Avalanche.

(1) If your are caught in an avalanche, call out so the other members of your patrol know to watch you as you are carried down the slope, and then keep your mouth closed to prevent ingestion of snow.

(2) Discard pack, skis and poles although this is much easier said than done. This gear tends to drag you underneath the surface of the moving debris.

(3) Assess best line of escape.

(4) Delay your departure, i.e., let as much of the avalanche pass you as possible.

(5) Try and work to the side. There will be less force of the avalanche at the edge of the flow.

(6) Try to swim out using a swimming and rolling action to stay on the surface of the snow.

(7) As you feel the snow slow down, thrust your arm or hand or any part of your body above the snow surface so that others can see it. You will probably be so disoriented that you won't know where the surface is so just guess and lunge.

(8) Before the snow comes to a rest, cup your arm or hand in front of your face to clear an air space. If possible, try to expand your chest during this time.

(9) If buried, stop fighting and relax to preserve oxygen. Remember, you are not suppose to panic!

d. Rescue Action. Use the Acronym STOP and GO. STOP means Stop, Think, Observe, Plan. GO means Go into action and Organize the rescuers. Do not panic! You are the victim's best chance of survival now.

(1) Watch the victim as he is carried down the slope. If the victim disappears under the moving snow, keep your eyes fixed on the mass of snow he was enveloped in, until it comes to a rest. The victim may be under the snow surface in that area.

(2) Also, using a ski pole, mark any position where he reappeared during his journey down the hill.

(3) Make a quick visual search of the area. Note any arms, legs, avalanche cord, and pieces of equipment, which appear and dig them out.

(4) At first, if nothing is apparent, make a quick surface search.

(5) If nothing is found, a more systematic search should be made from the bottom working up.

(6) If you again fail to find anything, your next step is to probe, which will be covered later.

(7) Stay on site and search. Almost all hope of a live rescue depends on you. Statistically, a victim has only about a 50% chance of survival if buried 30 minutes. The first 15 minutes are critical. Outside help cannot usually arrive fast enough.

**UNITED STATES MARINE CORPS**
Mountain Warfare Training Center
Bridgeport, California 93517-5001

FMST.07.02
08/21/01

**STUDENT HANDOUT**

# ENVIRONMENTAL HAZARDS II (SNOW STABILITY EVALUATION)

**TERMINAL LEARNING OBJECTIVE**. Given a unit in a cold weather environment and necessary equipment and supplies, assess winter environmental hazards, in accordance with the references. (FMST.07.02)

**ENABLING LEARNING OBJECTIVES**.

(1) Without the aid of references, select from a given list the description of the types of avalanche triggers, in accordance with the references. (FMST.07.02k)

(2) Without the aid of references, select from a given list the signs of instability, in accordance with the reference. (FMST.07.02l)

(3) Without the aid of references, select from a given list the signs of stability, in accordance with the references. (FMST.07.02m)

(4) Without the aid of references, conduct a snow pit analysis, in accordance with the references. (FMST.07.02n)

(5) Without the aid of references, conduct a rutschblock test on a slope, in accordance with the references. (FMST.07.02o)

**OUTLINE**.

1. **AVALANCHE TRIGGERS**. (FMST.07.02k) There are two types of triggers, natural and artificial.

a. Natural Triggers. These are not triggered directly by man or his equipment. A falling cornice, sluffing snow, stress change due to metamorphism in the snow pack, can all trigger avalanches.

**23**

b. Artificial Triggers. Man or his equipment triggers these. A skier passing, a mountaineer's weight, an explosive blast, a sonic boom, and the like commonly set off avalanches.

**NOTE:** An important fact is that artificial triggering leads to a far greater frequency of avalanches on a given path than if the path was left up to avalanche naturally.

2. **SIGNS OF INSTABILITY AND STABILITY**. To prevent ourselves from being the trigger, we need to formulate an opinion about the stability of the snow early on. Nature can provide some clues for us to use in determining the sensitivity of the snow pack.

a. Signs of Instability. (FMST.07.02l) Skilled avalanche hazard evaluation can be based upon a systematic decision-making process using signs of nature. You may experience these indicators by themselves or together. The following are signs of instability:

(1) Recent avalanche activity on similar slopes and small avalanches under foot.

(2) Booming. The audible collapse of the snow layers (normally kinetic growth metamorphism collapsing).

(3) Visible cracks shooting out from underfoot (severe tension in the snow pack).

(4) Sluffing debris is evidence of avalanche activity occurring.

(5) Sunballing, which is caused by rapid re-warming.

(6) Weather patterns

(a) Excessive snowfall, over 1 inch per hour for 24 hours or more.

(b) Heavy rain that warms and destroys the snow pack.

(c) Significant wind-loading causing leeward slopes to become overloaded.

(d) Long, cold, clear, calm period followed by heavy precipitation or wind-loading.

(e) Rapid temperature rise to near or above freezing after a long, cold period.

(f) Prolonged periods (more than 24 hours) of above freezing temperatures.

(7) Snow temperatures remaining at or below 25°F which slows down the settlement/strengthening process thus allowing unstable snow conditions to persist longer.

b. Signs of Stability. (FMST.07.02m) The following are signs of stability:

(1) Snow cones or settlement cones. These form around trees and other obstacles and indicate the snow around the object is settling.

(2) Creep and Glide. Creep is the internal deformation of the snow pack. Glide is slippage of the snow layer with respect to the ground. Evidence of these two properties on the snow pack is a ripple effect at the bottom of a slope. It is an indication that the snow is gaining equilibrium and strength through this type of settlement process.

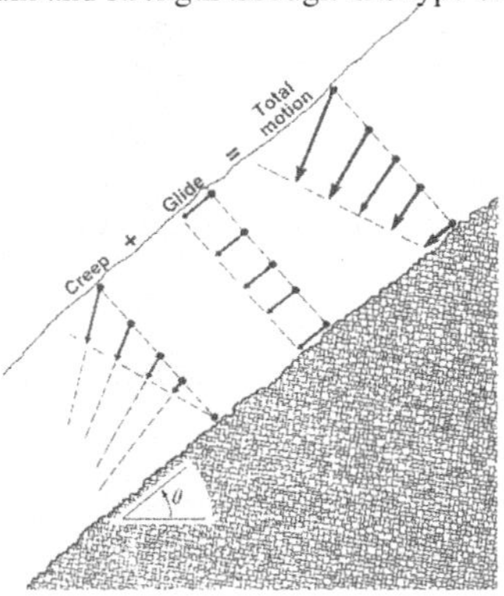


CREEP AND GLIDE

(3) Absence of Wind during Storms. This is indicated by snow accumulation on the trees.

(4) Snow Temperatures. Snow temperatures remaining between 25°F and 32°F ordinarily settles the snow rapidly, becoming denser and stronger because of the effects of equilibrium metamorphism.

3. **FIELD EXPEDIENT STABILITY TESTS**. Often, no single field test or observation will tell it all. You must piece together the story the snow pack is trying to tell by gathering up all available information. You will usually find that the various pieces of information back each other up and tell the same story. Field expedient stability tests are a good place to start gathering information, but keep in mind that they should be conducted on short slopes where no serious consequences would result.

(a) Small Steep Hills. The tester can sometimes get very useful feedback from a slope that is only a few feet high by jumping from the top onto the slope. Remember to take note of how they respond.

(b) Test Skiing. This is a stability test whereby a skier adds stress to the snow through his weight and/or by jumping and kicking. The tester can immediately observe the depth and type of weak layer that might have failed.

(1) When traversing uphill on skis and have just turned a corner, jump just below your uphill ski track and see if you can get a "piece of the pie" to break into blocks. This indicates that the snow may be cohesive enough to propagate a fracture.

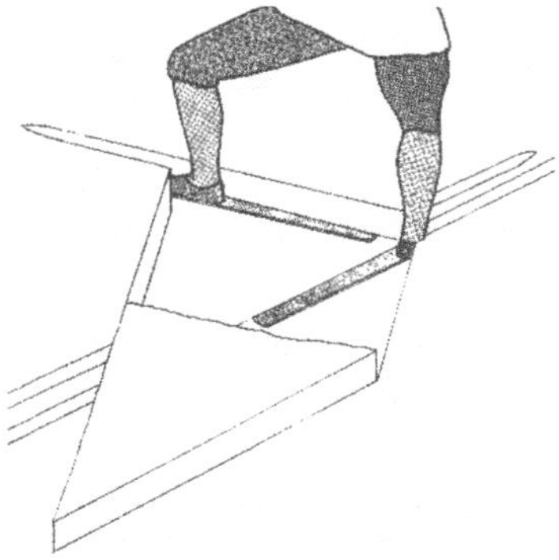

EXAMPLE OF A SKI TEST FOR STABILITY

(c) Ski Pole Test. This test takes only seconds and should be done often as you travel.

(1) Holding your ski pole at a right angle to the snow surface, gently push the pole into the snow pack.

(2) Feel for the relative hardness and the thickness of the layers.

(3) Be alert for well-consolidated layers that feel harder than underlying soft, weaker layers. This is a method to keep track of the depth and distribution of potential slabs.

(4) If the basket of the ski pole interferes with probing, use the handle of the ski pole to probe instead.

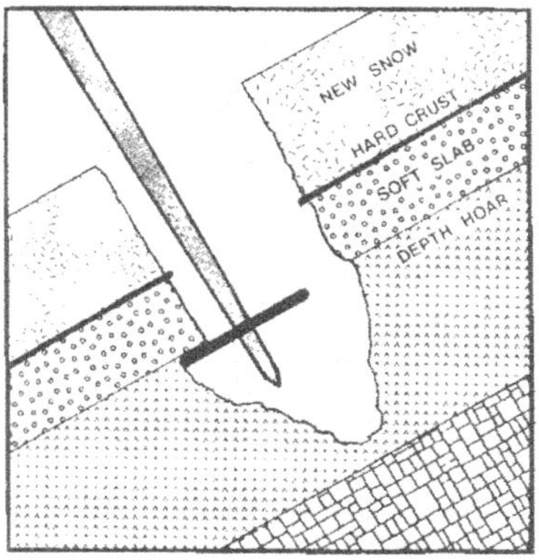


SKI POLE TEST

**NOTE**: One serious limitation of the ski pole test is that sometimes the weak layers are too thin to detect and it does not detect how well the layers are bonding together.

4. **SHEAR TESTS**. The principle objective of a shear test is to locate weak layers and interfaces. There are many different types of shear tests, but the two types we will discuss are the rutschblock and shovel/ski shear tests.

(a) Rutschblock Test. The rutschblock test involves loading a block of snow by a person in several stages. The following is the procedure for constructing the rutschblock.

(1) Construction

a. Select a site on a slope with the same slope angle and aspect as the slope that you are concerned with. Personnel conducting the test may be belayed.

b. Begin digging a pit approximately one-ski length in width and at least 4 to 5 feet deep.

c. From the ends of the pit, dig two narrow trenches uphill into the slope approximately the length of one ski pole. Ensure that the depths of the trenches are the same as the pit's depth.

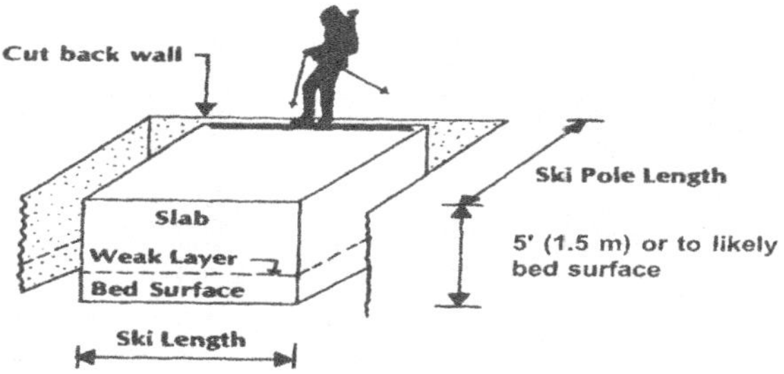


RUTSCHBLOCK TEST

d. Being very careful not to disturb the area of the rutschblock, use a snow saw or a knotted length of cordage and cut the back of the wall. This will isolate the snow block.

(2) Testing. Carefully ski to the side of the test site and approach the upper cut of the rectangular block diagonally from above. Once your skis are perpendicular to the cut on the uphill side, gently move on to the block. The test is conducted in stages contained in the following chart. When the block fails the chart will give you an idea how stable the slope is.

| STEP | REACTION | RESULT |
|---|---|---|
| One | Fails while excavating test site. | Extremely unstable |
| Two | Fails while approaching test site | Extremely unstable |
| Three | Fails while standing on shear block | Extremely unstable |
| Four | Fails while flexing your knees | Unstable |
| Five | Fails with one jump while wearing skis | Unstable |
| Six | Fails after repeated jumps with skis on | Relatively stable |
| Seven | Doesn't fail after repeated jumps with skis off | Stable or very stable |

(b) Shovel/Ski Shear Test. This is a quick method to obtain information of a location where weak layers are suspected, without involving a lot of digging.

(1) Construction.

a. Isolate a column in the uphill pit wall by cutting away the sides with a shovel or ski.

b. The width of this column as well as the depth cut into the pit should be approximately 12 inches. Ensure that the column is both vertical and smooth.

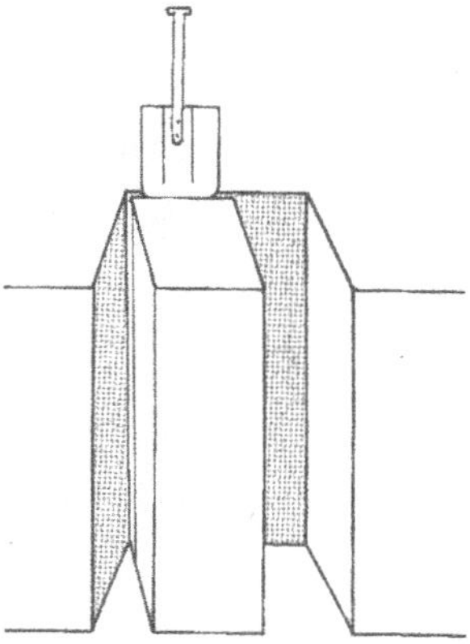

SHOVEL SHEAR TEST

(2) Testing. Insert a shovel/ski behind the column and exert steady pressure as you work your way down. Look for possible separation of the weak layers.

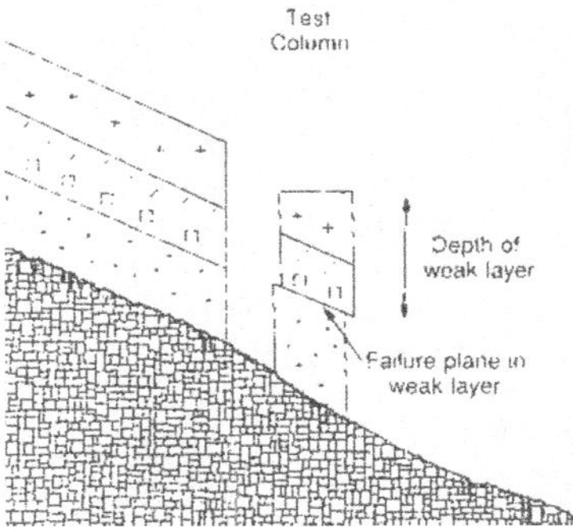


WEAK LAYER GIVING IN A SHOVEL SHEAR TEST

5. **SNOWPIT ANALYSIS**. In this section we will discuss how to analyze the snow pack for instabilities by identifying weak layers. Snow pit analysis can be extremely complex, but we will deal only with the basic observations.

a. Construction.

(1) Choose a location with the conditions similar to those you are trying to evaluate. They should at least be at a similar elevation, snow condition, slope angle and aspect as the slope you are concerned about.

(2) Dig a pit 4 to 5 feet deep and wide enough to work in. Be careful not to disturb the snow surface surrounding the uphill portion of the pit.

(3) With a shovel, smooth off the uphill pit wall and adjacent (side) wall. Ideally the adjacent wall should be shaded. These walls are where your tests will be conducted. It is important that they are smooth and vertical and that the snow above the uphill wall remains undisturbed.

b. Identifying Layers.

(1) Stratigraphy Test. Using a whiskbroom, paint brush, hat or mitten; lightly brush the sidewall of the pit with uniform strokes parallel to the snow surface. This will quickly transform the wall from a plain white surface into a layered mosaic of snow history. The raised or ridged surfaces indicate the harder, stronger layers that may be possible slabs or sliding surfaces. The indented surfaces reveal softer, weaker layers.

(2) Resistance Test. Insert a credit card, saw, or any straight edge into the top of the sidewall. Run the card down the wall, feeling the relative resistance of the layers and noting the boundaries of hard and soft layers. In helping to identify potential slab and weak layers, this test can help corroborate and expand upon the information gained from the Stratigraphy test.

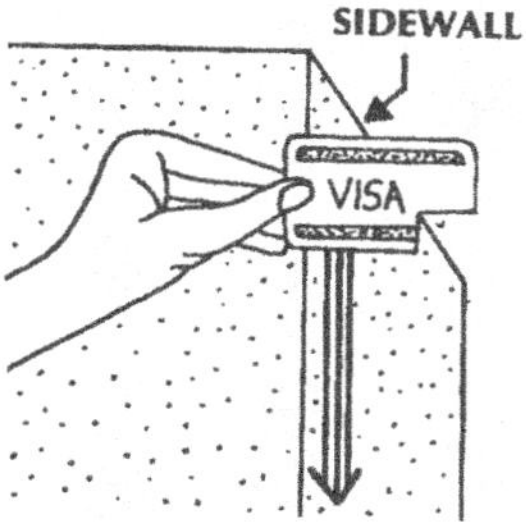


RESISTANCE TEST

(3) Hardness Test. Test the relative hardness of each layer by gently pushing your hand or fingers into the pit wall, applying approximately 10 lbs. of pressure. One layer might be so soft that you can easily push your whole fist into it while anothers might require a knife to penetrate it. An example of a potential unstable slab configuration would be a cohesive 1-finger hard layer resting on top of a less cohesive fist hard layer.

| | | |
|---|---|---|
| Very Soft | Fist (F) |  |
| Soft | Four fingers (4F) | 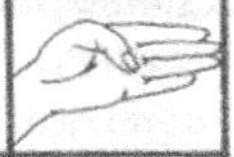 |
| Medium | One finger (1F) |  |
| Hard | Pencil (P) |  |
| Very Hard | Knife (K) |  |

HARDNESS TEST

**UNITED STATES MARINE CORPS**
Mountain Warfare Training Center
Bridgeport, California 93517-5001

LP
10/24/01

**STUDENT HANDOUT**

**AVALANCHE SEARCH ORGANIZATION**

**LESSON PURPOSE** The purpose of this class if to familiarize the medical support personnel with the avalanche search techniques so that they may be able to better support those directly involved in the search.

**OUTLINE**

1. **HASTY SEARCH.** A buried victim's life depends crucially on the action of survivors in the few minutes following an avalanche. Organized rescues are often slow getting to the scene due to the location and inaccessibility. Avalanche rescue is very similar to dealing with a drowning victim – "self-help is the key". When persons are caught in an avalanche, the survivors should react calmly and methodically. The following actions should be taken when witnessing an avalanche:

    a. Quickly look to make sure you are in a safe location.

    b. Immediately make note of the last seen point of a victim.

    c. Do a head count to determine who was caught in the avalanche. Also try to determine the last location of the missing victims when the avalanche occurred.

    d. Quickly assess the hazard of other possible avalanches. Post a guard (most likely the radio operator) at a safe spot. Make sure that all escape routes and warning signals are understood by all.

    e. Conduct a visual search of the deposit surface with concentration on the most likely burial areas. Look for parts of the victim and their equipment on the surface.

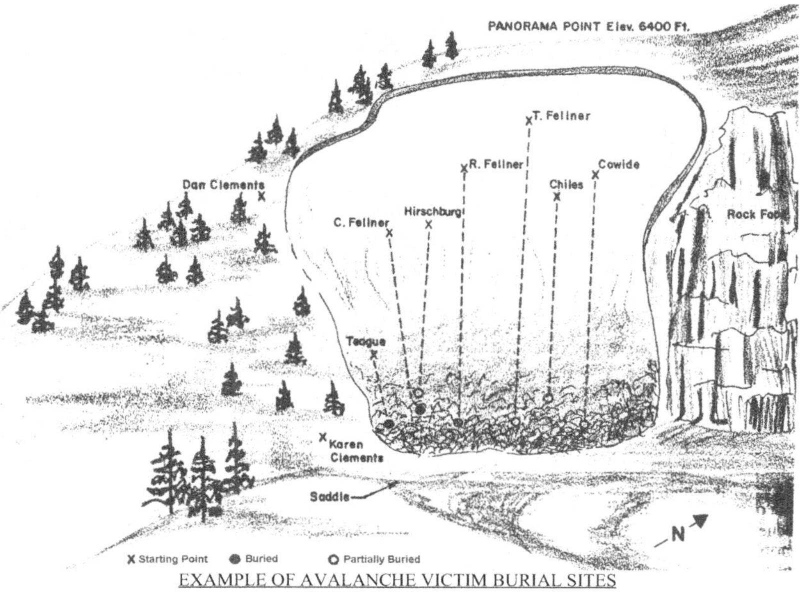


EXAMPLE OF AVALANCHE VICTIM BURIAL SITES

f. Draw a line from where the victim was caught, to the location of equipment and then of the victim's last seen point. The end of this line may point to the most likely burial site.

g. Have the radio operator alert the command of the situation to include location and number of victims involved. Request an avalanche search organization immediately.

h. Began searching the most likely burial areas. Simultaneously keep looking for signs of the victims and their equipment.

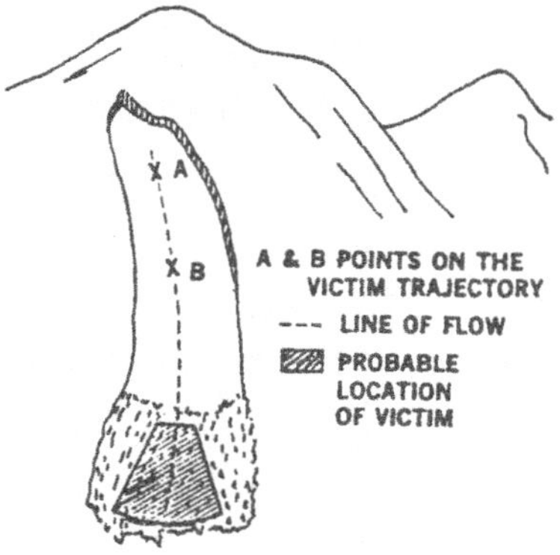


SUSPECTED LOCATION OF AVALANCHE VICTIM

i. Stop from time to time, call out to the victims and listen for voices of the buried persons.

j. Consider keeping packs on during the search so that shovels, probes and first-aid equipment are readily available.

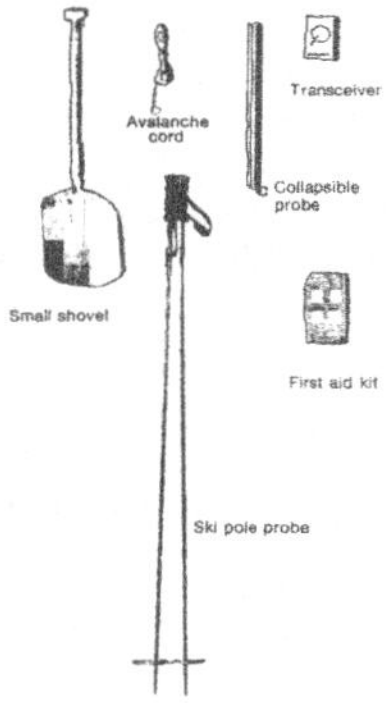


SEARCH AND SAFETY EQUIPMENT

k. Keep everyone involved in the search by probing and investigating most likely burial areas.

l. When victims are located, quickly dig them up and perform first aid as necessary.

m. If not all victims are located, mark the locations of clues found on the surface and spot probe around them.

n. Continue to spot probe most likely areas of burial to include areas behind trees, rocks, in depressions and in the run-out zone.

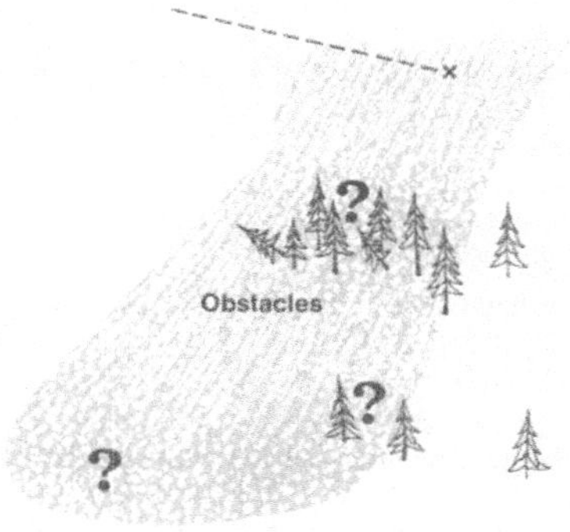


SUSPECTED SEARCH AREAS NEAR OBSTACLES AND RUN OUT ZONE

o. Organize a hasty probe line to probe the most likely burial areas and mark areas already probed with ski poles, skis, branches, etc. Techniques of probing will be discussed later in this chapter.

p. Try to keep the surface of the avalanche clean of food. This will prevent distraction of search dogs, if utilized.

q. Keep searching and probing until help arrives to take control over the search operation.

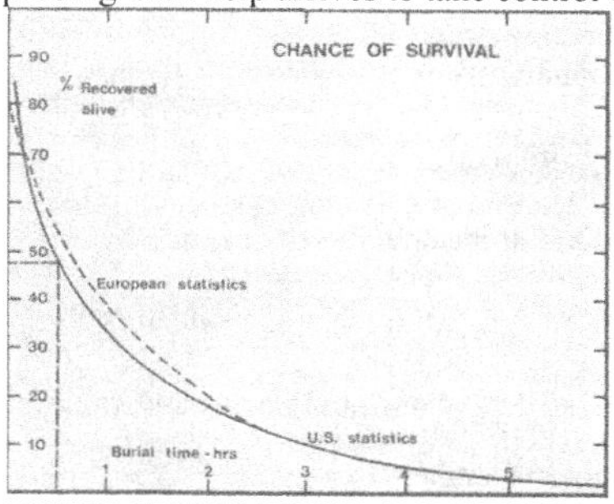


AMOUNT OF TIME BURIED AND SURVIVAL %

2. **AVALANCHE SEARCH ORGANIZATION**. A well-trained and equipped unit conducts an organized rescue. Normally this unit will be the unit of the victims. Once the commander is alerted that his Marines will be performing the search, he must collect his command element and proceed to the avalanche site.

a. Upon arriving at the site, the CO must make an estimate of the situation and make the following determinations, in order of priority:

(1) Evaluate the accident site.

(2) Post an avalanche guard and arrange for a warning signal, if not already established.

(3) Designate escape routes.

(4) Question witnesses and survivors at the scene about:

(a) How the accident happened.

(b) Persons buried.

(c) Locations of the unit's members when the avalanche occurred.

(d) Last seen point of the victims.

(e) Search efforts conducted so far.

(5) Provide care to the survivors.

(6) Determine most likely burial areas.

(7) Keep notes of actions and sketch a map of the avalanche site with location of clues.

(8) Locate a safe location for the Command Post.

(9) Locate a helicopter-landing zone.

(10) Consider the enemy situation.

b. Upon the company's arrival at the site, the CO must:

(1) Have the platoons store their equipment at a safe area away from the avalanche site.

(2) Inform platoon commanders about safety measures, the accident and action taken so far.

(3) Delegate tasks to each platoon, i.e. hasty search teams, avalanche guards, probe lines, etc.

(4) Have platoon commanders organize the probe lines.

(5) Consider equipment, food, and support that may be needed for a prolonged rescue.

(6) Keep the search organization focused.

3. **CONSIDERATIONS AND PROCEDURES FOR ASSIGNED TASKS**. As each platoon receives its task from the CO, there are certain considerations and procedures, which should be adhered to for each task.

a. Posting of avalanche sentries.

(1) Above natural anchors and starting zones.

(2) Sentries must prevent anyone from entering starting zone areas.

(3) Sentries must be in a position to observe any adjacent starting zones and prevent anyone from entering this zone.

(4) Sentries must be out of danger themselves.

(5) Sentries must be equipped with a signal device that will warn everyone about possible avalanche threats.

(6) The avalanche sentries may also serve as a security element for the search organization.

b. Establishing the Command Post.

(1) Set up warming tents for the searchers and victims, and prepare hot wets.

(2) Set up an aid station.

(3) Emergency medical sled(s) should be ready with sleeping bag, sled teams, and a corpsman.

(4) Stamp out and mark the LZ.

(5) The CP must be close enough to support the search yet far enough away from any existing avalanche hazards.

(6) Ensure that all radios are all on the same net and that communication is established to the next higher command.

(7) Provide guides to escort personnel from the road head to the accident site.

c. Hasty Search Teams.

(1) Search gullies and ravines, which could channelize a victim.

(2) Search uphill of catchment areas such as rock outcrops, trees, and benches in the slope and fallen logs.

d. Probe Lines.

(1) Snowshoes should not be worn on the probe line, as the debris of a hard slab avalanche will make snowshoe movement difficult.

(2) The ends of two adjacent probe lines must overlap by two men to insure that there is no gap between the lines.

(3) Since downhill probe lines can easily get out of alignment and probers tend to take overly large steps, the Probe Line Leaders must ensure that the probe lines remain aligned and in order.

(4) Overlap the flanks and run-out zones by at least 20 feet. Victims have been recovered from these areas after being shoved there by the force of the avalanche, even though the snow surface remains undisturbed.

(5) All areas searched must be marked to avoid confusion.

e. Other considerations.

(1) Dogs and individuals with transceivers will search independently of the probe line. However, they should have probers and shovelers readily available to uncover any possible strikes.

**NOTE**: The tactical situation may dictate how long the search may be conducted.

4. **PROBING**. There are two types of probes, the coarse probe and the fine probe. Each probe line will consist of two squads, and numerous markers and shovelers. The Platoon Sergeant will perform the duties as the Probe Line Leader. His job is to control the tempo of the line and ensures that the probers stay abreast.

a. Coarse Probe. The following steps will be taken for the coarse probe:

(1) With two squads on line, at close interval and at a designated area of search, each man will place the probe between his feet.

(a) The Probe Line Leader will be located behind each probe line.

(b) Also behind each probe line are the markers. Their mission is to place a mark where a strike has been indicated by a prober.

(c) Along with each marker is a team of shovelers whose job is to uncover the mark.

(2) The Probe Line Leader will give the command "DOWN PROBE", at that time the probes are then pushed down through each layer of the snow, being careful not to impale a victim if a strike is made.

(3) If a strike is made, the prober signals a marker to place a mark on the spot. The shovelers will then dig up this marked area. The line will never stop at a strike.

(4) The next command given is "UP PROBE", and all of the probes will be withdrawn from the snow.

(5) At the command "STEP", each man takes a 30-inch step and the process repeats.

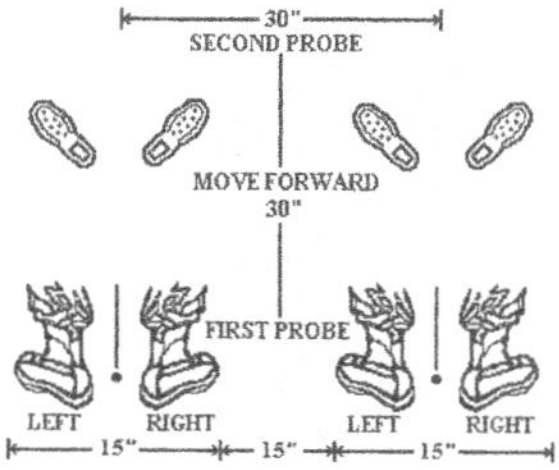


COARSE PROBE

b. Fine Probe. This is similar to a coarse probe with exception to the following:

(1) Probing is performed over the left, middle, and right foot

(2) A 15-inch step is taken rather than a 30-inch step.

(3) A fine probe is usually a body recovery and should only be started when all hope of a live recovery is exhausted.

(4) A fine probe search takes from four to five times longer than a coarse probe.

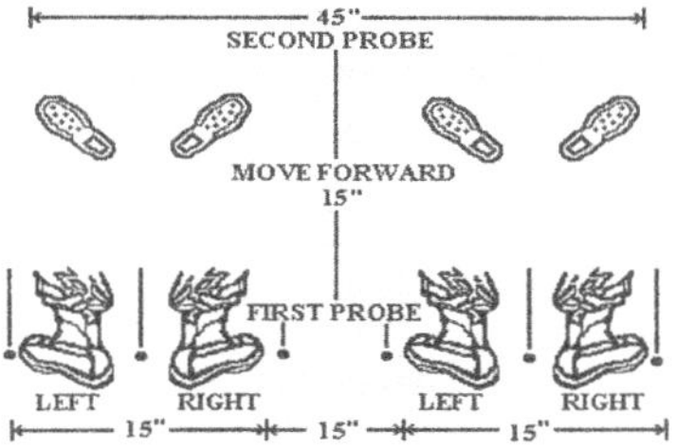


FINE PROBE

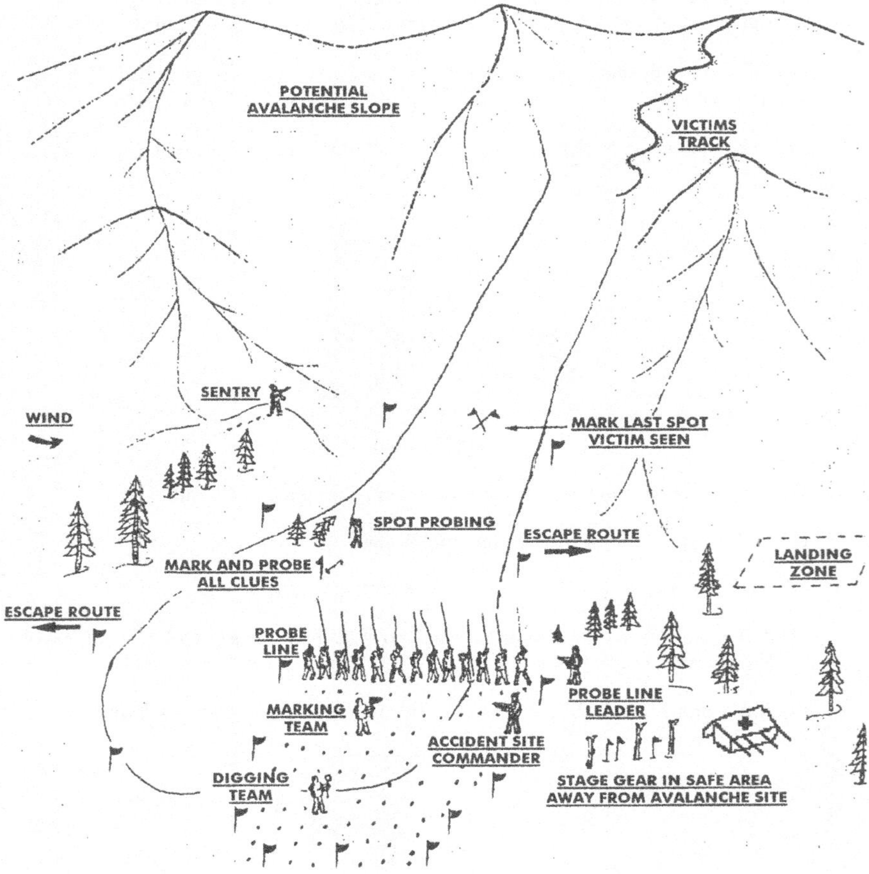


AVALANCHE SEARCH SITE

5. **SPECIALIZED SEARCHES**. These searches operate independently from other searchers.
5

a. Dogs. A probe line takes approximately four hours, but a dog takes about 20 minutes at 15 feet in depth. Searchers will allow the dog handlers to perform their duty without interference.

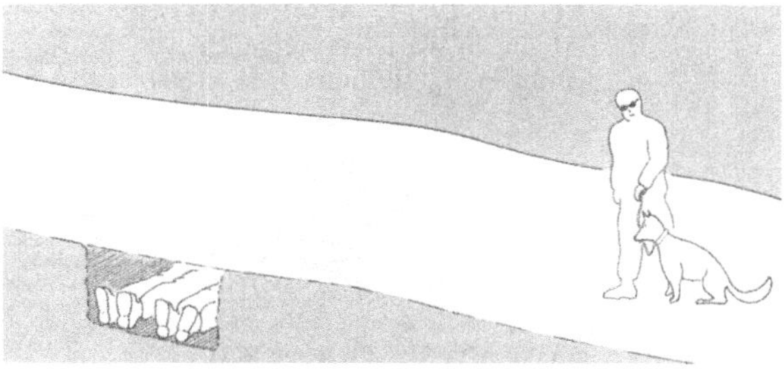

DOG SEARCH

b. Transceivers. This technique will be discussed in further detail in Chapter 12.

c. Sonar. Requires qualified operators as well as the obvious specialized equipment. Engineer units are normally outfitted with this equipment. If your unit has engineer support, then they should be assigned with this task.

**UNITED STATES MARINE CORPS**
Mountain Warfare Training Center
Bridgeport, California 93517-5001

LP
10/24/01

**STUDENT HANDOUT**

**AVALANCHE TRANSCEIVERS**

**LESSON PURPOSE** The purpose of this class is to familiarize the medical personnel with how an avalanche transceiver operates. There will also be a demonstration on how to locate a simulated avalanche causality using the avalanche transceiver.

**OUTLINE.**

1. **AVALANCHE TRANSCEIVERS**. They are very similar to a two-way radio, which emits an electromagnetic signal and can also receive signals that you can hear through either its built in speaker or an earphone.

   a. People who travel the backcountry cannot depend on probe lines for back up protection. They need a device for speedy detection regardless of the size of the avalanche.

      (1) At least once a season, practice drills are needed to breed familiarity with the transceivers.

      (2) Several systems are being marketed which have non-compatible frequencies. The solution to this problem is to buy a dual transceiver, which will have both the American 2.275 KHz and the European 457 KHz, which is what we utilize at MWTC.

      (3) The American frequency of 2.275 KHz is an audio frequency, which will not receive interference from radio, broadcasts and has a simple circuitry and less chance of malfunction. The main disadvantage is that the size of the antenna being small decreases the range.

(4) The European frequency of 457 KHz is a radio frequency that is affected by radio transmission interference. Its advantages are less power consumption and a longer range of 200 feet vice the American frequency at 100 feet.

**OPERATION OF THE TRANSCEIVERS**. There are many different brands of transceivers on the market. The two most commonly used transceivers at MWTC are the Pieps and the Ortovox.

a. Pieps

(1) To transmit:

(a) Slightly pull out the earpiece.

(b) Turn knob to the "1" position.

(c) Push the ear piece back in.

(d) The Pieps DF is now ready for transmitting.

(2) Search and receive

(a) Put earpiece into ear.

(b) Turn knob into "l" position at max volume.

(c) The Pieps DF is ready for receiving. Search the entire avalanche area until the signal is heard and move towards that direction. Upon getting closer to the victim, gradually lower the volume so you can home in on the victim's location.

(3) Wearing. To wear this device, place it around your waist and secure it with the fastening buckle.

**NOTE**: When drawing out the earpiece, the socket is extracted automatically and thus the Pieps DF is set to receive. If the socket is not pulled out, the earpiece was not snapped in properly. Push the earpiece firmly into the socket and pull it out all together. Turning the earpiece will make it easier to remove.

b. Ortovox

(1) Transmit

(a) First of all, before commencing the ski tour, put your Ortovox transceiver on correctly. Take the neck strap out of the covering case.

(b) Put the neck strap around your neck and insert the circular lock into the opening. Press this down against the spring and make a 1 /4 turn until the seam locks pointing to the arrow on the cover of the transceiver's case. Thus, the transceiver is automatically switched on.

(c) Check the battery condition by observing the test light on top of the transceiver case.

(d) Close the top flap of the carrying case.

(e) Check that the body strap is firmly wrapped around your chest.

(f) The Ortovox now transmits at the same time on both internationally used frequencies, 2.275 KHz and 457 KHz. You will hear the typical double signal of your Ortovox transceivers available on the market.

(g) Release it by using gentle pressure on the lock from the opening with a 1 /4 turn. Thus, the transceiver is automatically switched off. Release all straps and store your transceiver until your next skiing tour.

(2) Receiving

(a) When one or more companions are caught and buried by an avalanche, remember that the victim's life depends on you. DO NOT PANIC!!

(b) Take your Ortovox transceiver out of the carrying case and pull out the earphone with a binding, which is so fastened that it cannot be lost.

(c) Initially, all transceivers are set on "TRANSMIT". Now switch all transceivers to "RECEIVER". This is achieved by inserting the earphone plug into the socket on the bottom of the transceiver next to the battery lid; this will automatically switch it to receive. Make sure that all transceivers are set on receive since a single searcher with his transceiver set on transmit can frustrate the entire search.

(d) Place your earphone with the binding at the external ear and draw the cable behind the ear. After pulling the cable, slipping of the earphone is omitted.

c. Ramer

(1) Transmit. When the battery is inserted into the transceiver, the transmitter is automatically engaged.

(2) Receive

(a) Place earphone in receive/transmit jack, this will engage the receiver.

(b) The switch on the side is a loudness switch.

(c) To turn off, remove the battery.

## 2. DO'S AND DON'TS

a. Do’s

(1) Set transceiver to transmit at the beginning of your tour, as it could be difficult after you are caught in an avalanche.

(2) Use fresh alkaline batteries.

(3) Carry transmitting unit in side your clothing suspended by straps because a high-speed avalanche can remove a surprising amount of clothing and you don't want your transceiver to go with it.

(4) Carry an avalanche probe.

(5) Carry an avalanche shovel. Remember 4 to 6 minutes without oxygen and brain damage may occur.

(6) Store without batteries because of moisture and battery leakage.

(7) Still cross avalanche paths one man at a time because there will be competing signals in a multi-burial, making the locating of a single victim more difficult.

(8) Be careful when probing so you don't perforate the victim.

(9) Test all transceivers for transmit and receive before tour begins.

b. Don'ts

(1) Put the transceiver inside your pack because it will more than likely be the first thing you will get rid of.

(2) Wear quartz watches because of the tac-tac disturbance in the earphone.

(3) Remove your transceiver until the end of your ski tour.

(4) Change orientation of transceiver in space while moving. This "confuses" the transceiver.

(5) Carrying transceivers doesn't mean you can take added risks. Dead bodies have been recovered with working transceivers on them.

3. **SEARCHES WITH TRANSCEIVERS**

a. Carefully note and mark last seen point of the victim. If there is further danger of avalanches, post an avalanche guard and be prepared to switch back to transmit.

b. Be sure all transceivers are switched to receive. The leader must check.

c. Deploy line of searchers at a maximum of 30 meters apart at the last seen area and move down the slope.

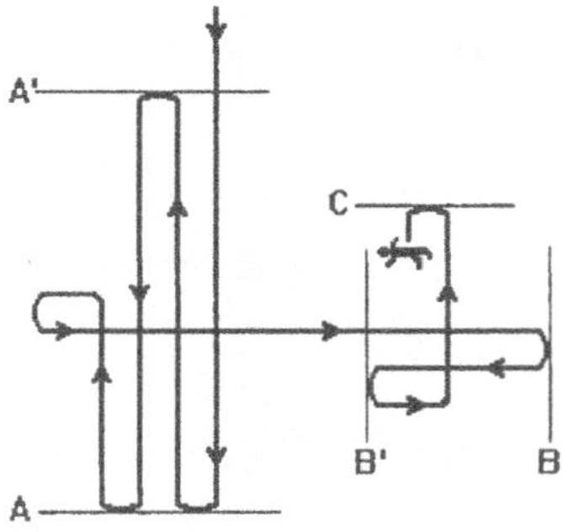

BRACKETING SEARCH

d. Volume control should be all the way up until the first signal is received.

e. Move in unison on line stopping every 10 paces and rotating the transceivers to check for signal.

f. When the first signal is heard, everyone is informed but the line should not break up.

(1) Bracketing to within 3 to 6 feet of the victim, the searchers probe to pinpoint the victim to avoid having to dig up too much snow and losing time.

(2) Depth of burial does not affect the signal range but does affect volume intensity.

**UNITED STATES MARINE CORPS**
Mountain Warfare Training Center
Bridgeport, California 93517-5001

FMST.07.03
10/24/01

**STUDENT HANDOUT**

**BIVOUAC ROUTINE**

**TERMINAL LEARNING OBJECTIVE**. Given a unit in a cold weather, mountainous environment, and necessary equipment and supplies, conduct a cold weather bivouac routine, in accordance with the references. (FMST.07.03)

**ENABLING LEARNING OBJECTIVES.**

(1) Without the aid of references, and given the scenario of being in snow covered mountainous terrain, select from a given list one area to avoid when selecting a bivouac site, in accordance with the references. (FMST.07.03a)

(2) Without the aid of references, select from a given list where to store the fuel bottle in accordance with the references. (FMST.07.03b)

**OUTLINE.**

1. **SITE SELECTION**. There are several factors to be considered in picking a good bivouac site. Higher Headquarters will assign you to a general sector, still leaving you with a lot of leeway in picking the specific site. Things to look for in that site, from most to least important are:

a. Not near or under a suspected avalanche site/run out zone. (FMST.07.03a) It is quite conceivable that in such an area, one enemy soldier with an explosive charge could wipe out an entire unit.

b. A good defensive position. In choosing between two good sites. Pick the one that is easily defended. Remember, *it is better to be miserably alive than comfortably dead.*

c. Be sure that the area is large enough to contain your whole unit. In mountainous terrain, it will be difficult to find an area large enough to accommodate a whole company. Usually, platoons will have to set up near a centralized point to company integrity but still establish their own bivouacs with a 360-degree defense. As always, camouflage and concealment is essential; it will take only one misplaced tent to give away your position.

d. In a forested area. There are several reasons for choosing a forested area:

(1) Natural cover and concealment under the trees for tents, vehicles, and tracks.

(2) Protection from the wind.

(3) Readily available firewood and defensive position construction material if needed.

(4) The trees can conceal and disperse smoke.

e. If a forested area is not available, then choose nearby depressions or knolls and dig down.

f. Near an adequate water supply. Snow can be melted, but this is a time and fuel consuming process and it may not be as clean as a running stream.

g. Generally, the best sites are on the leeward sides of mountains. Often just into or below the tree line offers an excellent site.

h. Off the valley floor. Cold air will settle during windless periods.

2. **ESTABLISHING THE BIVOUAC**. The order of establishing a bivouac is a very important consideration. If it is not properly adhered to, a unit will probably spend a lot of extra hours trying to establish it correctly. The following sequence should be second nature to a unit in a cold weather mountainous environment: security, track plan, defensive positions, living areas and specific use areas.

a. Security. Security is always essential. After a march in a cold weather environment, the tendencies will be to set up tents, rest and eat with no consideration for security or concealment of the bivouac site. While a unit is waiting to enter the bivouac site, they should be staged in a good defendable position, which will offer protection from the elements if possible. SECURITY MUST ALWAYS BE STRESSED.

b. Track Plan. While the unit leaders are doing their recon, they are going to be establishing a track plan. No unnecessary tracks are made. Good track discipline is going to be vital for the leaders so that the unit doesn't get confused and start making a major highway system around the bivouac site. A map study with a visual recon from afar will aid the leaders in knowing how they can set up their bivouac site to minimize confusion upon arrival.

(1) Jump-off Point. This is where the trail to the bivouac site meets the approach trail. The jump-off point must not be detectable by the enemy. It should also be well concealed by such things as large trees, rivers, boulders or other natural obstacles that will make it hard to detect. Generally when making the jump-off point, leave the main trail at a right angle, or leave it heading back toward the original direction of March for best deception. It should be covered by fire from the defensive position.

(2) Dummy Track. This is a deceptive trail that extends past the jump-off point on the main trail. Care should be used by the troops making this trail, so it fools the enemy into thinking that the unit has continued on the main trail. The main trail should extend well beyond the bivouac site to an area that can be used as a dummy bivouac site. When returning to the jump-off point, care should be exercised to ensure that your unit is not leaving marks to indicate that they were returning on this track. Some examples to aid in the return to the unit's position are:

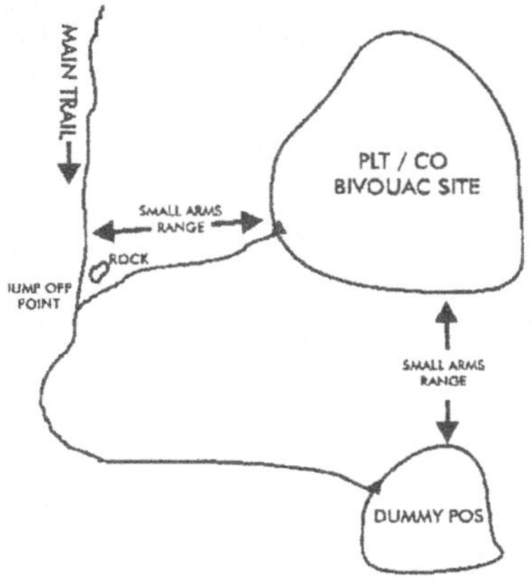

PLATOON / COMPANY TRACK PLAN

(a) If on skis, return without using your ski poles.

(b) Reverse snowshoes and walk with them on backwards. A string may be attached to the tails to make walking easier.

(c) Looping the dummy track, which will allow the men making the track to circle the track in front of the unit's actual defensive position, as well as provide the convenience to avoid retracing their tracks.

(3) Dummy Position. The end of the deceptive track can be made to look lived in. However, you don't want to alert the enemy to your presence in the area, so don't make it too obvious. If the enemy should find your trail, it is better that he should think you are in a dummy position, rather than being detected in the actual bivouac site.

(4) Track plan in the bivouac site. Every area in the bivouac site should be designated in the track plan.

(a) Central tracks should interlock everything so troops don't start making their own tracks.

(b) The defensive positions should be designated at this time.

(c) Before stamping out the tent sites, the leaders should have a good knowledge of the tent size. This will avoid having to enlarge the area by chopping trees, etc., or stamping an area for a tent that is to small and being tasked with recovering it.

(5) Camouflage and concealment of the track behind you. Camouflage is a continual process in the bivouac site. Here are some considerations for the track plan:

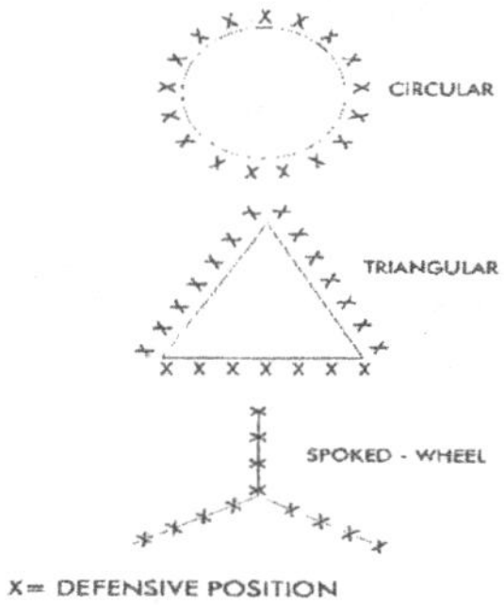


PLATOON DEFENSE

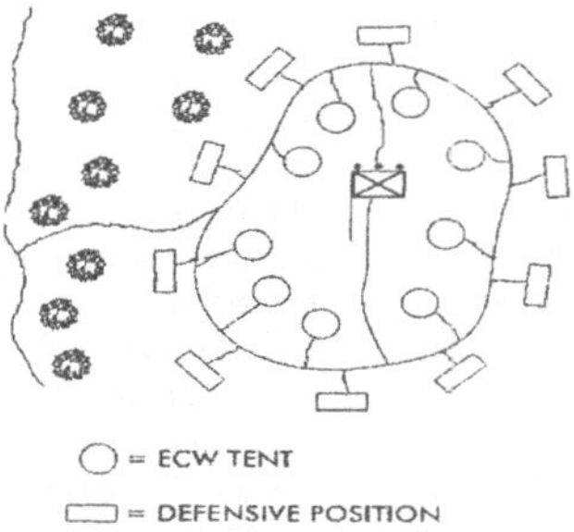


**PLATOON TRACK PLAN WITH DEFENSIVE POSITIONS**

(a) Tracks should meander as much as possible avoiding straight lines. Straight lines are easier to detect because there are few things in nature that are perfectly straight.

(b) A track that is covered by trees will also be harder to detect from above. Use tree cover as much as possible.

(c) Re-emphasize track discipline for everyone in the unit. One or two stray tracks may give away your whole position.

c. Defensive Positions. The tactical situation will dictate the type of defense established. Some of the options are: circular perimeter, triangular perimeter, or a spoked-wheel perimeter. Each type of perimeter has its pros and cons. The unit commander must weigh his options and choose a type best suited for his needs. Individual positions should be constructed in accordance with DEFENSIVE POSITIONS AND FIELD FORTIFICATIONS. Here are some things that must be considered when establishing your defensive positions:

(1) Security must be maintained while constructing the positions.

(2) Positions are constructed so they can cover the whole perimeter.

(3) Automatic weapons are positioned to cover the jump-off point, dummy position, and likely avenues of approach.

(4) Your defensive positions must be on the outside of the perimeter just beyond the range of the noises generated from inside the perimeter, so that the noises do not hinder the sentry from listening to his front. Approximately 30 meters is a good rule.

(5) All positions, i.e. fighting holes, tents, heads, etc., will be connected by communications trenches.

(a) The walls of the trenches will be constructed at an angle and the edges will be rounded off so that they do not cast shadows.

(b) These trenches should be chest to shoulder deep to protect you from incoming fire.

(c) They will be constructed so that they will be afforded the best camouflage and concealment.

(d) The trenches should be made in a zigzag pattern to avoid receiving fire down the long axis.

(6) At a minimum, 2 men per squad should stand arctic sentry duty at a time on the squad's sector of responsibility.

(a) Arctic sentry duty consists of a double staggered watch to ensure that the sentries are fresh and alert at all times. For example: Two men are in fighting positions at 0200 hours; one of these men came on post at 0130 and the other at 0200. The first man will be relieved at 0230 and the second man will be relieved at 0300. This allows one of the sentries to be fresh at all times.

(b) In extreme cold temperatures, a fire watch may also be needed. His duties are:

1. Maintain communication with the sentries through wire or other means.

2. Prepare a hot wet for the sentries upon their return from post.

3. Alert the others in case of danger.

4. Be a snow-watch during storms to prevent the tents from collapsing.

d. Living Areas. Once defensive positions are identified and manned, the Marines may start the construction of their living areas. These living areas must be clearly marked during the leader's recon and are connected to all other positions by comm. trenches. The actual construction and organization of these living areas will be discussed in depth later in this period of instruction.

e. Specific Use Areas. The final step in establishing the bivouac site is to designate and establish specific use areas.

(1) Head area. This will be centrally located but downwind of the living areas. It should not be so close to the living areas that people may get sick from it, nor so far away that people will not make the trek to use it in bad weather. If the tents are dispersed over a large area, more than one head area may be built. The head should be erected in

a relatively sheltered area out of the wind, or erect some ponchos/tarps to protect the user from the wind. Heads can be constructed as follows:

(a) The sleeve of an MRE case, when lined with plastic bags, makes an acceptable toilet. However, in wet conditions, the cardboard may get soggy and collapse under the weight of a Marine.

(b) Plastic buckets or boxes, also lined with plastic bags, make an excellent alternative.

(c) If establishing a long-term bivouac, lashing a sturdy pole in between two trees at about knee height can make a toilet. This pole can be used to sit on while defecating. A second pole can be lashed to the backside of the trees to provide a backrest.

(d) Urinals may be placed outside the individual tents for easy access. Measures should be taken to avoid urinating anywhere besides the designated "piss tree".

**NOTE**: Here at MWTC, all human feces will be double-bagged, clearly labeled as "human waste", And extracted down the mountain to the sanitation plant.

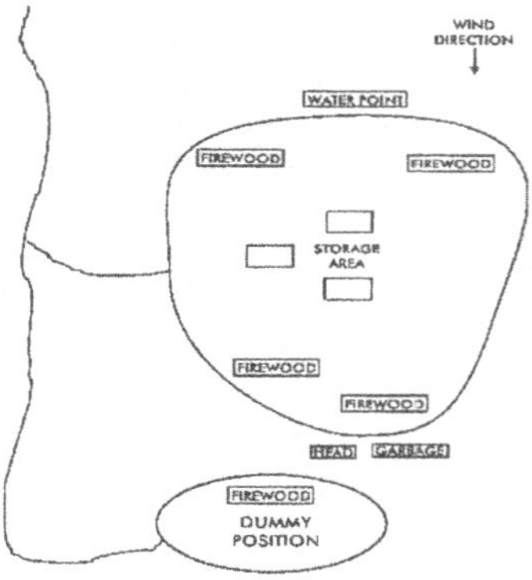


**SPECIFIC USE AREAS**

(2) The <u>water point</u> is the next site designated. If it is a stream, it will be the furthest point upstream. If no stream is available, then a large, clean, sheltered snow bank must be identified and marked off. It should be located upwind and as far from the head area as possible. If chemical or biological agents have been used at anytime in the past, the whole snow bank must be tested for contamination before use.

(3) The <u>garbage point</u> will be located next to your heads. All garbage will be retrograded without exception. In combat, the garbage could be disposed of in the dummy position.

(4) <u>Storage points for the unit</u> are also designated. These areas are inside the perimeter and fall under the control of the company gunny / police sergeant. These areas are for the unit's excess gear and equipment such as; vehicles, rations, fuel, ammunition, communications equipment, extra skis, etc. Some special considerations are:

(a) All gear will be protected from the elements to maintain serviceability.

(b) Safety precautions must be taken. For example: fuel should be stored at least 25 meters from any flame or explosives/ammunition.

(c) Points will be clearly marked to identify each point, and each man should know the location.

(d) All gear/equipment should be dug down and properly camouflaged and concealed.

(5) An area to gather firewood and building materials must be designated. Some key points to consider are:

(a) Spread you're cutting out so as not to give away your position by defoiling one part of the forest.

(b) Do your entire cutting during daylight hours so there are more natural noises to cover up your activities.

(c) You may consider cutting and gathering your wood from the dummy position.

3. **ESTABLISHING THE LIVING AREAS.** To maintain proper bivouac discipline and unit efficiency in a cold weather environment, there are many things that need to occur when establishing your bivouac.

a. Tent Site. To establish a tactical bivouac, the following steps should be taken:

(1) The tent site should, when possible, be located under overhanging tree limbs or near bushes. This provides anchor points for securing the tent, protection from the wind, and also helps conceal the tent from enemy observation.

(2) The tent site should be located at least 10 meters off of the main communication trench. This provides ample room for members of the tent to perform necessary functions (such as equipment maintenance) outside of their tent without blocking the access of the main trench line.

(3) When digging the tent down, ensure you pack down the snow to create a smooth, firm floor. This will help prevent restless nights.

(4) Ensure the pit you dig is large enough to allow you to walk around the exterior of the tent. This allows you to remove the snow that will build up on the roof of the tent during snowstorms.

(5) Position the tent in the pit with the entrance at the downwind side. This will help reduce the wind blowing inside the tent every time the door is opened and closed.

(6) Build a snow wall around the perimeter of the tent with the snow removed from the pit. This snow wall will help protect the tent from the wind, conceal it from enemy observation, and limit the amount of light that escapes from the tent when entering/exiting at night.

(7) The tent trench should be dug with a sharp bend in it. This will prevent the possibility of the enemy shooting down the trench's axis and into the tent.

(8) Offset the tent's entrance from the comm. trench. This will help enforce light discipline by reflecting the light from an opened tent door back into the tent vice down the length of the trench.

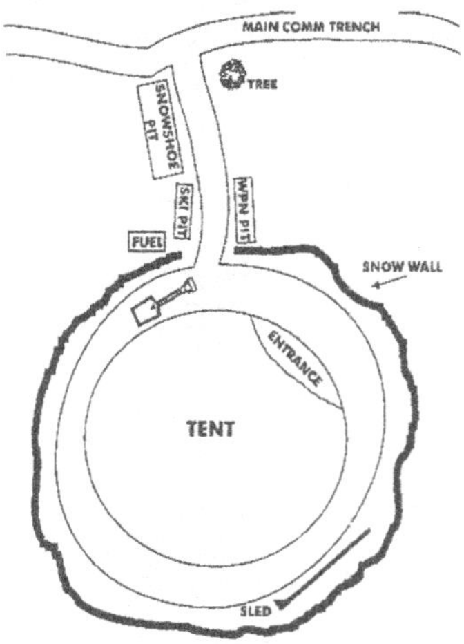


**ECW TENT LIVING AREA**

b. Organizing the Exterior of the Tent. A Marine unit hauls a lot of gear and equipment with them in a cold weather environment. Due to the size of the ECW tent, the majority of this gear must be left outside of the tent. To ensure no gear is lost or unaccounted for when displacing bivouac during periods of reduced visibility, a unit should have a set SOP as to where and how this excess gear is stored. The following is a recommendation as to how to store this gear.

(1) Ski Pit. Build a ski pit to the left of the tent entrance. The pit should be long enough to accommodate the skis while laying flat on the surface, and wide enough for all four pair of skis and poles. It only needs to be deep enough to allow the skis to be stored below the surface of the snow pack. Place pine boughs or branches on the floor of the pit to rest the skis on. Failure to place these branches will result in the skis actually freezing to the snow, forcing you to scrape the ice-buildup off before using the skis.

(2) Snowshoe Pit. Also on the left side of the entrance, build a pit for the team's snowshoes. This pit is constructed in the same manner as a ski pit.

(3) Weapons Pit. On the right side of the entrance, build a pit for the team's weapons (personal and crew served), and extra ammunitions. This pit is also constructed in the same manner as the ski pit. It is also necessary to cover these weapons and ammunitions with a poncho or some sort of tarp to protect them from the elements. Some unit commanders may prescribe that all weapons are kept inside the vestibule of the tent instead of being left outside. If this is the case, strict attention must be paid

to the effects of condensation as discussed in *EFFECTS OF COLD WEATHER ON INFANTRY WEAPONS AND OPTICS.* In extreme CW environments, the weapons will be stored outside regardless.

(4) Fuel Storage Area. (FMST.07.03b) On the left side of the tent, and at least 1 meter from the tent, designate an area to store extra cans and bottles of fuel. It is suggested that these cans/bottles are dummy-corded to a tree branch or tent line to facilitate locating them in the event of heavy snowfall. To prevent a tent fire, all refueling of stoves/lanterns must be done outside, away from the tent at the fuel storage area.

(5) Shovels. Ensure you leave your shovels near the door of the tent so you can keep your trenches free of new snowfall.

(6) Team Sled. Place all unused gear and equipment inside the sled, secure the cover, and place the sled on its side against the back of the tent.

(7) Piss Tree. Designate a tree or mound of snow as the tent team's piss tree. This tree should be about 5 meters or so from the entrance of the tent.

c. Organizing The Vestibule Area. Proper organization of the vestibule area of the tent is just as important as the organization of the exterior and interior of the tent. The following procedures should be followed:

(1) Cold Hole. Dig a rectangular shaped hole about 1 to 2 feet deep, in between the vestibule door and the tent door. This hole serves two purposes. First it provides a place to trap cold air and prevents it from drafting into your tent. Secondly, it provides a place to stand in when entering and exiting the tent.

(2) Packs. The team's packs should be left outside the main tent and stored in the vestibule. This creates more room inside the tent and prevents dragging excess snow inside the tent. All personal gear that is not being immediately used by the tent occupants should be stored inside their packs.

(3) Whisk Broom. A whiskbroom should be inside the vestibule. This broom is used to brush off the snow from the occupants prior to their entry into the tent. This prevents the snow from melting inside the tent, creating pools of water that soak personnel and their gear.

(4) Trash Bag. Place a trash bag inside the vestibule. The tent occupants need to place their trash in this bag as soon as it is created.

(5) Stoves / Cooking. All cooking and melting snow for water should take place in the vestibule. This is to prevent the inadvertent spilling of a pot of water or chow on the inside of the tent. Great care should be taken when lighting the stove inside the vestibule because a flare up could result in igniting the tent fly or tent body. The preferred method is to light the stove outside and then carry it into the vestibule. A squad stove will heat the vestibule and tent very quickly. However, it will also

consume all the oxygen in a sealed tent, resulting in asphyxiation. To prevent this, leave the top of the door unzipped about 8 to 12 inches.

d. Organizing the Interior of the Tent. Four Marines in an ECW tent can easily be compared to sardines in a can. In order to prevent temper flare-ups from the cramped living conditions, the following guidelines should be followed:

(1) Sleeping Arrangements. The occupant's sleeping bags and mats should be laid out in accordance with the MARINE CORPS COLD WEATHER INFANTRY KIT.

(2) Individual Gear. All individual gear that is not stored in the individual's pack should be kept neat, orderly, and staged in the individual's sleeping area. Do not allow your gear to invade into your fellow occupant's "space". When gear is not being used, it should be placed inside the individual's pack or WP bag to prevent it from getting wet.

(3) Drying Wet Gear. In a cold weather environment, it is inevitable that Marines will get their gear wet. The ECW tent has a mesh drying rack in the roof of the tent. All wet gear should be placed on the rack to dry. However, depending on the amount of gear, the occupants may have to take turns drying their gear. Too much gear on the rack will hinder the flow of air, thus slowing down the drying process. Excessively wet gear should be wrung out outside to prevent excessive dripping on the occupants and their gear. A lit stove can enhance the drying of clothes immensely. Ensure that there is proper ventilation to prevent asphyxiation.

(4) Lighting. To provide adequate lighting inside the ECW tent, the occupants can hang flashlights or chemlights from the drying rack buckles. Candles are not recommended for use inside the tent due to the fire hazard they create. Small lanterns work very well when hung from the drying rack. Lanterns, however, are a high fire hazard, and should be lit outside and brought into the tent once glowing.

(5) Ventilation. The interior of the ECW tent quickly becomes rancid with the scent of unbathed Marines, dirty socks, wet gear, etc. To help alleviate this problem, open up all doors and escape hatches on a regular basis. To prevent asphyxiation, always leave the bottom of the door unzipped for 8 to 12 to facilitate the flow of fresh air.

(6) Temperature. Temperatures inside the tent should be kept comfortably cool. This helps to conserve fuel and also helps maintain the acclimatization of personnel.

4. **GENERAL TIPS FOR LIVING IN THE COLD**. The following tips may help you adapt to living in the cold, not just survive it.

a. Before going to sleep at night, eat a hot meal and drink a hot wet. Putting warm fluids/food into your body keeps you warm longer. Also, keep some snacks handy to eat during the night to maintain body heat. Remember, food is fuel.

b. Wear the minimum amount of clothing in the sleeping bag as the tactical situation permits.

c. Canteens filled with warm water and placed in your sleeping bag make excellent hot water bottles that pre-heat your bag. Ensure the canteen caps are tightly secured or you may wake up with a wet sleeping bag. This also prevents your water from freezing over night.

d. Fill your thermos with a hot wet every night before going to sleep. This will provide you with a hot wet in the event you have to "bug-out" in the middle of the night.

e. Always keep your gear neat, orderly and packed in the event you have to displace suddenly.

f. Take care of your personal hygiene needs (shaving, sponge bath, etc.) at night prior to going to sleep. This will reduce the removal of natural skin oils that help prevent CW injuries, i.e. windburn, sunburn, frostbite, etc.

g. Keep all battery-operated equipment (flashlights, hand-held radios, etc.) inside the sleeping bag with you if feasible. Leaving this gear laying on the tent floor over night will kill the batteries.

h. The five-gallon water jug should be stored upside down to prevent the freezing of the water at the pouring point. Insulation covers are available through the supply system.

i. Handle all fuel outside the tent and always use contact gloves. In extreme cold temperatures, fuel spilled on unprotected skin will freeze the skin tissue almost immediately.

j. Whenever there is a stove or flame source inside a tent, maintain a fire watch.

k. Damp clothes (socks, glove liners, etc.) may be placed in the sleeping bag overnight to dry them. Do not put extremely wet clothes in the sleeping bags, as this will only make the bag wet.

l. Have a sponge or rags handy to mop up spills or excess condensation.

**UNITED STATES MARINE CORPS**
Mountain Warfare Training Center
Bridgeport, California 93517-5001

FMST.07.14

10/24/01

**STUDENT HANDOUT**

# WILDERNESS ORTHOPAEDIC/ TRAUMA INJURIES

**TERMINAL LEARNING OBJECTIVE:** Given a casualty in a mountainous environment, and necessary equipment and supplies, manage common orthopedic injuries, in accordance with the references. (FMST.07.14)

**EABLING LEARNING OBJECTIVES:**

(1) Without the aid of references, select from a given list the phases of clearing a cervical spine in the field in accordance with the references. (FMST.07.14a)

(2) Without the aid of references, select from a given list, the correct treatment for an open book pelvic fracture in accordance with the references. (FMST.07.14b)

(3) Without the aid of references, select from a given list, six considerations for orthopedic patient evacuation in accordance with the references. (FMST.07.14c)

OUTLINE

1. INTRODUCTION.

Orthopedic injuries occuring in the wilderness are not significantly different from those suffered in garrison. Bones break, ligaments are torn and strained, and joints are dislocated in both settings. The key to preparing for and managing wilderness orthopedic injuries is understanding that the provider will have significant limitations with regard to diagnostic tools and initial management resources at his/her disposal. The personal risks each expedition member is exposed to can rise dramatically if the team is needed to transport a victim out of the back-country. Determination of a casualty's mobility becomes a major consideration for the wilderness medical provider. Assessment of the terrain, creative use of available natural resources, and effective leadership are all necessary components of caring for orthopedic injuries in the wilderness.

2. GENERAL: Making the diagnosis.

The wilderness "doc", whatever his/her training background, needs to have good physical exam skills. NO X-RAY and NO LABS will be available in the back-country. These basic tools that we depend on, in garrison, will not in the field. The provider will need to rely on his/her observation, palpation, and auscultation skills to make a diagnosis. This person will also need to use the patient, if possible, to uncover the likely injury by taking a thorough history which includes careful review of the mechanism of injury.

A. HISTORY: Critical information can be gained from talking to your patient.

- Mechanism of Injury - Determine:
  1. High vs. low-velocity accident
  2. Direction of force
  3. Nature of activity
  4. Loss of consciousness

- Ask the casualty:
  1. Did they hear a pop, snap, crack, or breaking sound?
  2. How long ago did the injury occur?
  3. Can they move the injured body part?
  4. Was there swelling within 15 minutes of injury or did it take longer?
  5. Was there immediate disability or was there a gradual loss of function?
  6. Was there immediate discoloration of a joint?

- Other important information:
  1. Evaluate the environment. (i.e. Hot/Cold, Wet/Dry)
  2. Allergies.
  3. Medications currently being taken.
  4. Previous injuries/surgeries to the site in question.

B. PHYSICAL EXAMINATION

Your powers of observation become crucial.

- Look for:
  1. Swelling.
  2. Discoloration and bruising.
  3. Obvious deformity/angulation.
  4. Open wounds. (with or without protruding bone fragments)
  5. Differences right vs. left.
  6. Muscle spasm surrounding injured site.

7. Point tenderness.
8. Crepitus.

- Beware:

Consider the environment while assessing your patient
- Don't undress the casualty in cold!
- Protect your patient from the elements. (Sun, Rain, Cold)
- Talor your physical exam to meet the constraints of the environment.
    * palpate under clothing
    * visualize one region at a time, then redress
    * set up temporary shelter from the elements

3. Head Injuries:

a. A blow to the head may lead to increased intracranial pressure (ICP) or intracranial bleeding neither of which are managable in the wilderness. The job of the medical personnel is to diferentiate between a serious, life threatening injury and one which is minor.

1. Minor injury: No loss of consiousness (LOC), or LOC of less than 15 seconds with immediate return to full alertness. The casualty can not be on medications which increase risk of bleeding or have a history of bleeding disorders. Patient may be monitored every two hours for mental status changes, lethargy, irritibility, persistent nausea and vomiting, changes in speech or visual changes.

2. Serious Injury: LOC greater than 15 seconds, and/or persisitent confusion or memory loss; signs or symptoms of increased ICP: Debilitating headache, mental status changes, persistent nausea and vomiting, appearance of clear fluid in external auditory canal, Battle sign, raccoon eyes, seizures, or return LOC.

3. Field treatment of serious head injury: Suspect injury of C-spine, manage airway-be able to clear vomit, elevate head 30 degrees. Evacuate to treatment facility ASAP.

4. Spinal Injuries.

A. Cervical Spine

1. High Risk Activities:

C-spine injuries in the wilderness usually occur after either a fall from a significant height or, especially in the winter environment, a high-velocity accident. Common winter activities predisposing participants to C-spine injuries are skiing, snow-boarding, and snow-mobiling.

2. Anatomy:

The cervical spine consists of seven cervical vertebrae interposed between the base of the skull and the thoracic spine. The cervical spine has a great deal of mobility to allow maximal range of motion for the skull. This increased mobility comes at a high price; this portion of the spine is less stable than the rest of the vertebral column, and hence more subject to injury. The C-Spine is stabilized by three longitudinal ligaments: The anterior longitudinal ligament (ALL) runs longitudinally along the anterior surface of the vertebral bodies. It is broad and very strong and helps to prevent hyper-extension of the head and neck. The posterior longitudinal ligament (PLL) also runs longitudinally, this time along the posterior surface of the vertebral bodies, within the vertebral canal itself. This ligament is relatively narrow and somewhat weaker than the ALL, and helps to prevent hyper-flexion at the neck. Third is the supraspinous ligament, connecting the spinous processes. This, too, is a strong ligament, acting to inhibit hyper-flexion.

The peripheral nerves innervating the muscles and the sensory nerves of the upper extremities originate from the cervical portion of the spinal cord and these are some of the nerves likely to be affected in the event of a C-Spine injury. Hence, it is important to assess the neurologic status of the upper extremities if a cervical injury is suspected.

3. C-Spine Injury Statistics:

-Most common injury: Flexion injury at C5/6
(so look for a deficit in the C6 distribution)

-28% of C-Spine injuries have another spinal
fracture associated with it.
(so examine entire spine)

-10% of Head injuries/Facial fractures also have a
C-Spine injury, especially if there was LOC.

4. Remember:

a. Head injury -> Assume C-Spine injury.
b. Perform complete Neurovascular exam
   -Motor, Sensory, Reflexes, Pulses, Babinski

B. Clear a C-Spine.

When faced with a possible C-Spine injury in the Wilderness, clearing it should be considered only in extreme situations. However, there are situations which demand that every effort be made to make a casualty ambulatory, because the movement of the individual would demand herculean effort from the rest of the team, potentially placing more individuals at risk of injury or prolonging exposure to extreme environmental conditions.

Clearing a C-Spine in the field should follow a three-phased assessment. If the patient fails the assessment at any one of the phases, then the team is obligated to maintain spinal precautions and transport the patient. The phases are outlined below, and expanded upon in the lecture.

PHASE ONE:

-ensure no alteration in the consciousness of the patient is present
-maintain in-line traction
-perform complete Neurovascular exam (focusing on the upper extremities)
-palpate spinous processes one-by-one while asking casualty whether palpation causes pain (supraspinous ligament)
-distinguish between pain over spinous process and muscle soreness associated with paraspinal musculature.
-only if palpation was completely pain-free can you move on to ...

PHASE TWO:

-continue to maintain in-line traction
-palpate anteriorly left and right of the trachea along the anterior vertebral bodies (anterior longitudinal ligament)
-note bony tenderness upon palpation from deep under the angle of the jaw to the clavicle
-again, distinguish between tenderness of the sternocleidomastoid muscle and actual vertebral pain
-only if palpation was pain-free can you progress to...

PHASE THREE:

-while loosely maintaining in-line traction, have patient move head through active range of motion

*first flexion/extension
*followed by rotational movement

-stop test at the first indication of pain with movement

Note: If you are able to move the patient's head through a full range of motion without pain, you have effectively cleared his C-Spine. Have the patient resume activity slowly and cautiously and evaluate any other complaints noted.

B. Thoracic and Lumbar Spine:

-Look for a T-L Spine injury with calcaneal fractures approximtely 10% of patients with a calcaneal fracture will also have a concomitant lumbar fracture.

C. Pelvis:

-Think HEMORRHAGE/SHOCK if faced with a Pelvic Fracture
*Up to 6 liters of blood loss possible – (internal iliac areteries)*
Place gentle constricting wrap around pelvis if open book fracture is Suspected.

4. FRACTURES.

Some common wilderness fractures to prepare for:

| | |
|---|---|
| a. Metacarpal/Phalangeal | -the hands are your means of interfacing with the environment, sometimes it's a forceful meeting |
| b. Distal Radius | -the most common wilderness fracture seen |
| c. Scaphoid | -check for snuffbox tenderness, pain with axial load to thumb |
| d. Lunate | -also common, pain dorsally at base of 3rd MC |
| f. Clavicle | -need good NV exam, check for pneumothorax |
| g. Long bone | -Radius/Ulna: if proximal, check elbow<br>-Humerus: radial nerve runs in spiral groove<br>-Tib/Fib: think compartment syndrome- Fib fx may ambulate with cane<br>-Femur: think Hemorrhage, think traction |
| f. Ankle | -commonly a fracture/dislocation, check NV |
| g. Hip | - Leg will typically be externally rotated |

REMEMBER:
- Always think HEMORRHAGE with Long Bone fractures
- Always perform NEUROVASCULAR Exam before and after treatment
- Always SPLINT AND PAD for transport
- Indications for REDUCTION are: NV deficit , severe angulation, severe pain, if angulation prediposes to open fx or makes transport difficult.
- OPEN FRACTURES: Have eight hours to get to surgery. Field Tx- Gently wash off with betadine/ iodine solution to get off obvious dirt- don't scrub. Wrap in sterile gauze. Don't place exposed bone back under skin. Give antibiotics: Augmentin, 2nd or 3rd generation cephalosporin, a quinolone or tetracycline.

THE REASONS WE PLACE FEMUR FRACTURES UNDER TRACTION.

1. Re-establishing length tightens fascia and tamponades bleeding.
2. Dramatic pain relief.
3. Helps prevent open fractures.
4. Helps reduce secondary soft tissue damage.
5. Increases ease and safety of transport.

SOME BASIC RULES OF EXTREMITY SPLINTING:

- Splint all fractures before moving the casualty, unless imminent danger.
- Splint all fractures as they are found, unless severe angulation complicates transport or is causing neurovascular deficits.
- Splint to include the joint above and the joint below.
- Shape and measure splints using uninjured extremity, then splint injury in the position of function.

5. DISLOCATIONS.

a. Rapid Diagnosis and Reduction is imperative, if evacuation time is > 1-2 hours.

- Easier to reduce immediately after injury. (muscle spasm)
- Makes transport much easier. (increased patient comfort)
- Dramatic pain relief.
- Early reduction reduces risk of Neurovascular injury.
- Reduction could make difference between an ambulatory vs. litter patient
  *the safety of the entire expedition could be placed at risk*

b. Signs and Symptoms: Remember, NO X-RAYS!

- Decreased Range of Motion.
- Obvious Deformity. (compare right vs. left)
- Usually NO Crepitus.
- Typical Posture.

c. Reduction Techniques:

| | |
|---|---|
| - Phalanges | Linear traction, buddy tape. Can not reduce volar displacement of 1st phalanyx. |
| - Shoulder | Bedsheet method- counter traction<br>Prone with weights method<br>External rotation. |
| - Patella | Extension with medially directed pressure. |

| | |
|---|---|
| - Ankle | Linear traction with knee flexion. |
| - Hip | Anchor pelvis, hip flexed at 30 degrees, upward traction. |
| - Sternoclavicular | Roll between scapulae, forceful downward pressure on both shoulders simultaneously |
| -Knee | Linear traction, beware of popliteal vascular injury |

REMEMBER: Analgesia and Sedation should be utilized when available before attempting joint or fracture reduction, unless multiple traumatic injuries are present.

6. EVACUATION CRITERIA:

Conditions demanding expeditious evacuation:

1. Suspected cervical, thoracic, and/or lumbar spine injuries.
2. Pelvic injury with instability and/or significant blood loss.
3. Any open fractures. (8 hour window, antibiotics early)
4. Compartment syndromes. ( pain with passive motion, pallor, pulseless, paresthesia)
5. Hip or Knee dislocations.
6. Any other injury the medical provider feels he/she is not prepared to manage.

7. OTHER TRAUMA.

| | |
|---|---|
| Heavy bleeding: | Usually can be treated by direct pressure, if a tourniquet is required,release pressure every five minutes to reassess. |
| Contusion: | Apply ice 1st 48 hours which will provide pain relief and limit expansion of hematoma. |
| Lacerations and avulsions: | High pressure irrigation with clean water, may remove debris with flame treated forceps. Use steri-strips for minor clean lacs. Keep dirty or puncture wounds open and give antibiotics. |
| Animal Bites: | Copious irrigation, and antibiotic;, do not close. Consider rabies if bite was unprovoked and animal was acting strange. (bats, raccoons, skunks, canines, and felines) Has never been seen in livestock, rabbits, squirrels, rats or mice. Most common zoonotic source in North America is the skunk. |

| | |
|---|---|
| Severed part: | Can be reattached up to 24 hours after injury if the cut is clean. Gently clean part, wrap in gauze and keep cool without freezing. |

**UNITED STATES MARINE CORPS**
Mountain Warfare Training Center
Bridgeport, California 93517-5001

LP
10/25/01

**STUDENT HANDOUT**

# COLD WEATHER MOUNTAIN LEADERSHIP CHALLENGES

**PURPOSE**. The purpose of this period of instruction is to emphasize the vital role of leadership in the conduct of successful operations and to promote among leaders at all levels an understanding of the problems common to units operating in a cold weather mountainous environment. This lesson relates to all of the training that you will receive here at MWTC.

**OUTLINE**

1. **POSITIVE LEADERSHIP AND THE RIGHT ATTITUDE**. Leadership must be by example. At first, conditions tend to be frightening. Marines will find themselves up against many problems they have never met before and the environment will always be there to remind them that they can become a casualty if they make mistakes. Aggressive leadership, which cheerfully meets and overcomes the challenges of the environment, is essential to mission accomplishment. There will be two enemies to contend with the enemy soldier and the environment itself. The first step towards defeating these enemies is getting your Marines in the right mental attitude. The leader must maintain a positive attitude towards the mission, his Marines, and the equipment they have to carry out the job. You can be defeated psychologically, if you are not aware of the symptoms of a poorly motivated unit. The following are some tips on how to defeat such problems when they begin to get on the minds of your Marines.

    a. If your Marines withdraw into a shell or become moody and depressed, get them involved in conversations with each other.

    b. If your Marines find it hard to remember things they have been taught, show patience and review orders, drills, and SOP'S. Keep their minds busy.

    c. Tempers normally flare up during this type of training, so expect and are prepared to deal with it when it comes. Maintain your sense of humor, lead by example, and don't let unanticipated problems get the best of you.

d. Be alert for individuals who will place their own physical comfort ahead of their assigned duties. Remind them that their mission as Marines is to fight and to do so successfully requires that weapons and equipment be maintained in working order.

e. Don't accept excuses for not carrying out an order. Mountainous training all too often becomes a camping trip. Leaders must challenge their Marines to train, as they would fight.

2. **LEADERSHIP PROBLEM PECULIAR TO COLD WEATHER/MOUNTAIN OPERATIONS**

a. Cocoon-like Existence. Many men, when bundled up in successive layers of clothing and with their head covered by a hood, tend to withdraw within themselves and to assume what has been termed a "cocoon-like existence". When heavily clothed, the individual's hearing and field of vision are greatly restricted and he tends to become oblivious to his surroundings. His mental processes become sluggish and although he looks, he does not see. Leaders must recognize and overcome these symptoms. Additionally, the leader needs to watch for the growth of lethargy within himself and must be able to prevent it. He must always appear alert to his men and prevent them from sinking into a state of cocoon-like existence. The remedy is simple and basic: ACTIVITY. Throw the hood back and engage in physical activity. Although the remedy is simple, the recognition of the condition requires leadership.

b. Individual and Group Hibernation. This problem is again a manifestation of withdrawal from the in environment. It is generally recognized by a tendency of individuals to seek the comfort of sleeping bags, and by the group remaining in tents or other shelter at the neglect of their duties. In extreme cases, guard and security measures may be jeopardized. Again, the remedy is simple: ACTIVITY. The leader must ensure that all personnel remain alert and active. Rigid insistence upon proper execution of all military duties and the prompt and proper performance of the many group "chores" is essential.

c. Personal Contact and Communications. It is essential that each individual and group be kept informed of what is happening. Due to the normal deadening of the senses, a man left alone may quickly become oblivious to his surroundings, lose his sense of direction, his concern for his unit, and in extreme cases, for himself. He may become like a sheep and merely follows along, not knowing or caring whether his unit is advancing or withdrawing. Each commander must take strong measures to ensure that small unit leaders keep their subordinates informed. This is particularly true of the company commanders keeping their platoon commanders informed, of platoon commanders informing their squad leaders, and the squad leaders informing their men. General information is of value, but the greatest importance must be placed on matters of immediate concern and interest to the individual. The chain of command must be rigidly followed and leaders must see that no man is left uninformed as to his immediate surrounding and situation.

d. Time/Distance Factors. Mountain operations doctrine recommends that tactical commanders be given every opportunity to exploit local situations and take the initiative when the opportunity is presented. Because of the increased amount of time involved in a movement and the additional time required to accomplish even simple tasks, deviation from tactical plans is difficult. Tactical plans are developed after a thorough reconnaissance and detailed estimate of the situation. Sufficient flexibility is allowed for each subordinate leader to use his initiative and ingenuity in accomplishing his mission. Time lags are compensated for by timely issuance of warning orders, by anticipating charges in the tactical situation, and the early issuance of frag orders. Recognition of time/distance factors is the key to successful tactical operations in cold weather/mountainous regions.

e. Conservation of Energy. Two environments must be overcome in mountainous regions; one created by the enemy and the second created by the climate and terrain. The climatic environment must not be permitted to sap the energy of the unit to a point where it can no longer cope with the enemy. The leader must be in superior physical condition. He cannot expend the additional energy required by his concern for his men and still have the necessary energy to lead and direct his unit in combat. He must remember that there are seldom any tired units, just *TIRED COMMANDERS!*

**UNITED STATES MARINE CORPS**
Mountain Warfare Training Center
Bridgeport, California 93517-5001

FMST.07.12
10/25/01

**STUDENT HANDOUT**

**SUBMERSION INCIDENTS**

**TERMINAL LEARNING OBJECTIVE.** Given a simulated casualty, treat submersion incident victims, in accordance with the references. (FMST.07.12)

**ENABLING LEARNING OBJECTIVES.**

(1) Without the aid of references, define in writing submersion incident, in accordance with the references. (FMST.07.12a)

(2) Without the aid of references, select from a given list the definition of secondary drowning, in accordance with the references. (FMST.07.12b)

(3) Without the aid of references, select from a given list the respiratory/cardiovascular symptoms exhibited by a submersion incident casualty, in accordance with the references. (FMST.07.12c)

(4) Without the aid of references, select from a given list the correct field management priorities of a submersion victim, in accordance with the references. (FMST.07.12d)

(5) Without the aid of references, select from a given list three of the six favorable prognostic signs post submersion incident, in accordance with the references. (FMST.07.12e)

(6) Without the aid of references, select from a given list three of the four unfavorable prognostic signs post submersion incident, in accordance with the references. (FMST.07.12f)

1. **TERMINOLOGY.**

a. Definition. (FMST.07.12a) A submersion incident is the medical problem that occurs after a casualty has been submersed underwater.

(1) *Drowning* refers to death by suffocation following submersion in water (less than 24 hours).

(2) *Near drowning / secondary drowning* (FMST.07.12b) refers to submersion in water with at least temporary survival (greater than 24 hours).

2. **PATHOPHYSIOLOGY**

a. Sequence of events:

(1) Panic initially, followed by a violent struggle.

(2) Gulping and swallowing air and water to avoid aspiration.

(3) Breath holding until hypoxia leads to unconsciousness.

(4) Once consciousness is lost, the gag reflex relaxes and passive influx of water into the lungs occurs. A small percentage of victims (10-15%) have significant laryngospasm, which prevents any appreciable volume of water to enter the lungs. This is referred to as "dry drowning".

(5) Aspiration is more likely in immersion victims who enter very cold water (10$^{O}$C). An involuntary gasp reflex occurs, with a marked increase in respiratory rate and depth. This individual is at increased risk of aspiration, and subsequent drowning.

b. Submersion in very cold water will quickly result in hypothermia.

(1) Rapid induction of total body hypothermia (core temperatures of 89.6$^{O}$F or less) may be protective in submersion incidents. Significant decreases in oxygen requirements result, especially in the case of the brain itself. Marked decreases in cerebral blood flow and oxygen demand have been demonstrated (see Hypothermia lesson). This protective effect of hypothermia has been used to explain the many cases of complete recovery after extended periods of submersion, always in cold water.

(2) The "Mammalian Diving Reflex" slows the heart rate, shunts blood to the brain and closes the airway when more primitive marine mammals are exposed to extremely cold water. General consensus holds that this reflex is not a significant response in human beings. However, it has been found to be most pimounced in newborn human infants. It has also been speculated that this "Reflex" may be more significant in

those individuals who have survived extended submersion times. The issue has not been resolved to date.

c. Fatalities in immersion victims associated with hypothermia are primarily due to three known mechanisms.

   (1) Immersion syndrome. Sudden death due to cardiac arrest secondary to massive vagal stimulation. Usually associated with sudden total body exposure to frigid water.

   (2) Excessive fatigue and confusion leading to fatal errors of judgment.

   (3) Direct hypothermic effects. Thermal conductivity in water is approximately 25 times that of air. Despite the evidence showing hypothermia's protective effects, the complications associated with hypothermia (cardiac arrhythmias, CNS depression, loss of motor control) can lead to submersion and drowning.

d. Some authorities emphasize distinguishing between "wet" and "dry" drownings. In wet drowning, aspiration of water into the lung occurs. (85-90% of casualties). In dry drowning, laryngospasm prevents aspiration of water (10-15% of cases). Ultimately, the end result is the same: obstruction to respiration, oxygen depletion, and asphyxiation. The morbidity and mortality associated with the two types of drowning are not significantly different.

e. Fresh-water and salt-water drownings are somewhat different.

   (1) In fresh-water aspiration, the osmotic difference between hypotonic fresh-water and the relatively hypertonic blood leads to net water movement <u>into</u> the vasculature. The risk of hemolysis is great. This flow across the alveolar membrane also leads to "washout" of surfactant, which leads to collapse of the alveolar spaces. Hypoxemia follows.

   (2) In seawater aspiration, the osmotic shift is reversed, leading to fluid outpouring from the blood stream flooding the alveoli and the interstitium. This leads to decreased lung compliance as well as direct damage to pulmonary capillaries. Again, hypoxemia results.

   (3) Hypoxemia is the end result in both types of drowning. No significant difference in morbidity in survivors has been shown between fresh-water and salt-water drownings. The only bodies of water shown to have a worsening effect on prognosis are the hyper-saline bodies, like the Dead Sea and the Great Salt Lake. Otherwise, prognosis is similar.

f. Hypoxemia is the primary insult after submersion. Efforts to correct this problem should be vigorous and sustained. All clinically dead submersion casualties should receive CPR, or ACLS if available, immediately upon removal from the water, regardless of water temperature or duration of submersion. Since the overwhelming majority of naturally occurring water bodies in the world have temperatures below 92°F, all submersion victims should be aggressively as treated case reports of survival from submersion up to 66 minutes have been documented hypothermia casualties. Remember: No one is dead until they are warm and dead.

g. In the case of prolonged submersion or significant aspiration, pulmonary derangements occur rapidly. In lesser insults, the onset of symptoms may be delayed for as long as 24 hours. (FMST.07.12c) Physiological changes include:

(1) *Cardiovascular.* Casualties develop respiratory acidosis and hypoxemia, which may trigger dysrhythmias. Electrolyte changes are not usually significant in survivors of submersion incidents and should correct once spontaneous respirations are re-established.

(2) *CNS.* C-Spine injuries are common in submersion incidents, most commonly a result of shallow-water diving. Injury patterns include wedge, teardrop and burst fractures of the lower C-Spine, frequently resulting in tetraplegia. Conclusive head injuries also occur. Drug and alcohol abuse should always be considered.

(3) *Metabolic.* Respiratory acidosis is immediate problem. Acid/base derangement should correct after resuscitation of the casualty without specific intervention.

3. **SURVIVAL TECHNIQUES IN COLD WATER.**

a. Swimming motion increases heat loss more than it increases heat production. The direct hypothermic effects of cold-water immersion make it impossible to swim for long periods of time. The average swimmer is unlikely to swim more than a kilometer in 10$^{O}$C water. Swimming is advised only when a rescue vehicle is nearby, or the distance to shore is not great.

b. Treading Water and Drown Proofing.

(1) Treading water results in reduced survival time due to increased heat loss secondary to increased activity.

(2) "Drown Proofing" is a technique whereby the swimmer relaxes in the water, floating facedown, arms spread, allowing his face to submerge between breaths. "Drown proofing" is the least desirable survival technique in cold water since the head and neck, high heat loss areas, are maximally exposed.

c. Personal Floatation Device (PFD). Without the aid of the PFD, the swimmer caught in cold water is forced to use less desirable methods to stay afloat.

(1) If wearing a PFD, the individual can utilize the "Heat Escape Lessening Posture" (HELP). This position minimizes the exposure of the swimmer's high heat loss areas because the arms are folded across the chest and pressed to the sides, the knees are drawn up to the chest and the legs are crossed at the ankles. Unfortunately, this position is fatiguing and becomes difficult to hold.

(2) If there is more than one person in the water, the "Huddle" position should be used. Similar to a football huddle, the chest, abdomen and groin area should be pressed together. Children or injured individuals should be placed in the middle of the huddle, since they become hypothermic more quickly.

d. Increasing Survival Time:

(1) Try to enter the water in a lifeboat or raft.

(2) Locate and don PFD as quickly as possible.

(3) Wear several layers of clothes; especially protect high heat loss areas.

(4) If you must enter the water, enter gradually rather than jumping or diving.

(5) If afloat in the water, avoid unnecessary movement. Assume the HELP or Huddle position.

(6) Don't try to swim.

(7) Remember the situation is not hopeless. Have the will to SURVIVE!

4. **RESCUE TECHNIQUES.**

a. Reaching the Casualty. The basic order for water rescue is: Reach and Pull, Throw, Tow and, as a last resort, Go.

(1) *Reach and pull.* When the casualty is responsive and close to the shore, begin by holding out an object for him to grab, and then pull him from the water. Floating line is considered the best choice, but if it is not available, a tree branch, fishing pole, oar or other object will do as well.

(2) *Throw.* Should the casualty be alert, but too far away for you to reach and pull throw an object that will float to them. A PFD or ring buoy is best, if available. The primary course of action is to throw anything that will float, and to do it as quickly and as accurately as possible.

(3) *Tow.* Once the conscious casualty has an object to hold onto, try to find a way to tow him to shore. From a secure position, tow the casualty ashore.

(4) *Go.* Swim to the casualty as a last resort. You should be a strong swimmer, trained in water rescue, in order to avoid the dangers associated with this form of rescue. If the casualty is unconscious or unresponsive, you must go to the casualty.

5. **SIGNS AND SYMPTOMS.**

a. History of Incident. Several questions need to be answered, as they will help determine the approach to the casualty.

(1) When a person is found floating face down on the water surface or is brought to the surface by a rescuer, there usually is little question about the cause of unconsciousness.

(2) When near drowning is unwitnessed or occurs after a dive or other water activity (surfing), a head injury and a C-spine injury should be presumed, and C-Spine precautions should be instituted.

(3) Determine how long the casualty was submerged. This helps determine prognosis.

(4) How cold was the water? Very cold water (less than $70^{O}F/21^{O}C$) may be protective, giving rescuers additional time to begin resuscitation.

(5) What was the casualty's health prior to submersion? Drowning could have been secondary to alcohol ingestion, cardiac arrest or other illness/injury.

b. Casualties may be symptomatic or asymptomatic at the scene:

(1) Respiratory signs and symptoms may include dyspnea, tachypnea (SOB), sore throat, substernal burning, pleuritic chest pain, persistent cough, frothy sputum, cyanosis, rales, rhonchi and wheezes.

(2) Cardiovascular signs may include tachycardia and hypotension.

(3) Neurological findings may include anxiety, restlessness, lethargy, confusion, convulsions, incontinence, hyporeflexia and coma.

(4) Abdominal distention is common as is hypothermia.

6. **FIELD MANAGEMENT.**

a. The scene of a submersion incident is frequently chaotic and first consideration should be the safety of the rescuers, then retrieval of the casualty from the water.

b. Priorities: (FMST.07.12d)

(1) Establish an airway, breathing and circulation (ABC's). A trained rescuer can do rescue breathing while the casualty is still in the water.

(2) Observe C-Spine precautions, since cervical spine and head injuries are common, especially in diving accidents.

(3) Begin CPR as soon as possible, if necessary, when the casualty reaches the shore, or suitable rescue platform. When the victim is pulled ashore lay him or her horizontal to the shoreline. If the head or feet are placed down hill it encourages blood to pool in that area of the body, and is counter productive to CPR.

(A.) The symptomatic patient should be given high-flow O2 via a non-rebreathing facemask.

(B.) IV access should be established, if available.

(4) Medevac ASAP.

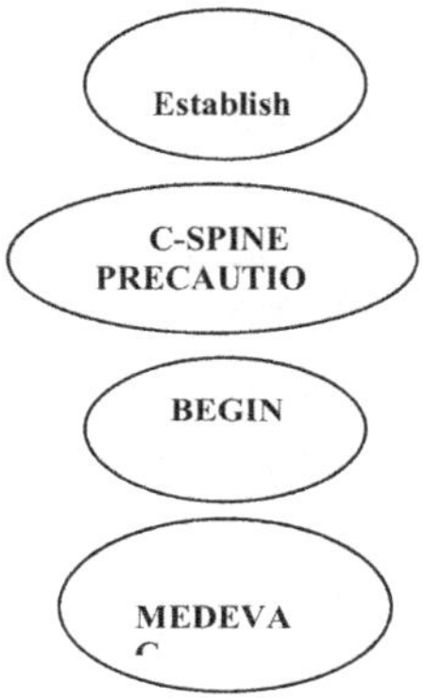


c. All victims of submersion incidents will be observed in a medical facility for a minimum of 6-8 hours. It is not uncommon for asymptomatic victims to develop respiratory failure within that time frame.

d. Vomiting is common and should be anticipated by turning the patient's head and having suction ready, if available. Frequently, the cause of vomiting is gastric distention.

(1) The American Heart Association recommends that gastric distention be relieved only when it interferes with CPR.

(2) "Breaking" the patient, i.e., placing the patient in the prone position and lifting upwards at the abdomen can achieve gastric drainage.

(3) Another method is to turn the patient to a lateral recumbent position and compress the abdomen with the palm of one hand.

(4) The preferred method of gastric drainage is to pass an NG tube and relieve distention with gentle suction.

(5) Gastric drainage should not be attempted without suction equipment available, due to the risk of aspirating vomitus.

7. **PROGNOSTIC PREDICTORS.**

a. 0 - 2 "Unfavorable Prognostic Indicators" yield a greater than 90% survival rate (see above).

b. 3 or more "Unfavorable Prognostic Indicators" yields a less than 5% survival rate.

c. Favorable prognostic signs: (FMST.07.12e).

(1) Casualty alert.

(2) Basic or advanced life support received at the scene.

(3) Good response to initial resuscitation.

(4) Hypothermic.

(5) Older child or adult.

(6) Brief submersion time.

d. Unfavorable prognostic signs: (FMST.07.12f).

(1) Age under three years.

(2) Fixed dilated pupils.

(3) Submersed five minutes or longer.

(4) No resuscitation attempts for the first ten minutes.

(5) PH less than 7.10.

(6) Cardio-pulmonary arrest as initial presentation.

(7) Resuscitation time greater than 30 minutes.

e. Conn's Criteria: Based on the status of the patient 1-hour after initial resuscitation.

| | |
|---|---|
| (1) A - Awake | All survive (99+%) |
| (2) B - Blunted | Most survive (95+%) |

(3) C - Comatose

| | |
|---|---|
| C1 - Flexor Response | Good survival with ICU care (80+%) |
| C2 - Extensor Response | Good survival with ICU care (80+%) |
| C3 - Flaccid | Poor survival |
| C4 - Arrest | Dismal survival |

**UNITED STATES MARINE CORPS**
Mountain Warfare Training Center
Bridgeport, California 93517-5001

FMST.07.19
10/25/01

**STUDENT HANDOUT**

**REQUIREMENTS FOR SURVIVAL**

**TERMINAL LEARNING OBJECTIVE**. In a winter mountainous environment, apply the requirements for survival, in accordance with the references. (FMST.07.19)

**ENABLING LEARNING OBJECTIVES**.

(1) Without the aid of references, and given the acronym "SURVIVAL", describe in writing the acronym "SURVIVAL", in accordance with the references. (FMST.07.19a)

(2) Without the aid of references and from a given list, choose the four priorities of work in a survival situation, in accordance with the references. (FMST.07.19b)

**OUTLINE**.

**1. REQUIREMENTS FOR SURVIVAL**

a. This mental "mind-set" is important in many ways. We usually call it the "will to survive" although you might call it "attitude" just as well. This basically means that, if you do not have the right attitude, you may still not survive.

b. A guideline that can assist you is the acronym " SURVIVAL". (FMST.07.19a)

(1) Size up.

(a) Size up the situation.

1. Conceal yourself from the enemy.

2. Use your senses to hear, smell, and see to determine and consider what is developing on the battlefield before you make your survival plan.

**30**

(b) Size up your surroundings.

1. Determine the rhythm or pattern of the area.

2. Note animal and bird noises and their movement.

3. Note enemy traffic and civilian movement.

(c) Size up your physical condition.

1. Check your wounds and give yourself first aid.

2. Take care to prevent further bodily harm.

3. Evaluate the condition of your self and unit prior to developing survival plan.

(d) Size up your equipment.

1. Consider how available equipment may affect survival senses; tailor accordingly.

(2) **U**ndue haste makes waste.

(a) Plan your moves so that you can move out quickly without endangering yourself if the enemy is near.

(3) **R**emember where you are.

(a) If you have a map, spot your location and relate it to the surrounding terrain.

(b) Pay close attention to where you are and where you are going. Constantly orient yourself.

(c) Try to determine, at a minimum, how your location relates to the following:

1. The location of enemy units and controlled areas.

2. The location of friendly units and controlled areas.

3. The location of local water sources.

4. Areas that will provide good cover and concealment.

(4) **V**anquish fear and panic.

(a) Realistic and challenging training builds self-confidence and confidence for a unit's leadership.

(b) The feeling of fear and panic will be present. The survivor must control these feelings.

(5) Improvise and Improve.

(a) Use tools designed for one purpose for other applications.

(b) Use objects around you for different needs. (I.e. use a rock for a hammer)

(6) Value living.

(a) Place a high value on living.

(b) Refuse to give into the problem and obstacles that face you.

(c) Draw strength from individuals that rise to the occasion.

(7) Act like the natives.

(a) Observe the people in the area to determine their daily eating, sleeping, and drinking routines.

(b) Observe animal life in the area to help you find sources of food and water.

**NOTE:** Remember that animal reactions can reveal your presence to the enemy. Animals cannot serve as an absolute guide to what you can eat and drink.

(8) Live by your wits, but for now, learn basic skills.

(a) Practice basic survival skills during all training programs and exercises.

2. **STRESS**. Stress has many positive benefits. Stress provides us with challenges: it gives us chances to learn about our values and strengths. Too much stress leads to distress. While many of these signs may not be self-identified, it remains critical that all survivors remain attentive to each other's signs of distress. Listed are a few common signs of distress found when faced with too much stress:

a. Difficulty in making decisions (do not confuse this sign for a symptom of hypothermia).

b. Angry outbursts.

c. Forgetfulness.

d. Low energy level.

e. Constant worrying.

f. Propensity for mistakes.

g. Thoughts about death or suicide.

h. Trouble getting along with others.

i. Withdrawing from others.

j. Hiding from responsibilities.

k. Carelessness.

3. **SURVIVAL STRESSORS**. Any event can lead to stress. Often, stressful events occur simultaneously. These events are not stress, but they produce it and are called "stressors". In response to a stressor, the body prepares either to "fight or flight". Stressors add up. Anticipating stressors and developing strategies to cope with them are the two ingredients in the effective management of stress. It is essential that the survivor be aware of the types of stressors they will encounter.

   a. Injury, Illness, or Death. Injury, illness, and death are real possibilities a survivor has to face. Perhaps nothing is more stressful than being alone in an unfamiliar environment where you could die from hostile action, an accident, or from eating something lethal. Illness and injury can also add to stress by limiting your ability to maneuver, get food and drink, find shelter, and defend yourself.

   b. Uncertainty and Lack of Control. Some people have trouble operating in settings where everything is not clear-cut. The only guarantee in a survival situation is that nothing is guaranteed. This uncertainty and lack of control also add to the stress of being ill, injured, or killed.

   c. Environment. A survivor will have to contend with the stressors of weather, terrain, and the variety of creatures inhabiting an area. Heat, cold, rain, winds, snow, mountains, insects, and animals are just a few of the challenges awaiting the Marine working to survive. Depending on how a survivor handles the stress of environment, his surroundings can be either a source of food and protection or a cause of extreme discomfort leading to injury, illness, or death.

   d. Hunger and Thirst. Without food and water a person will weaken and eventually die. Getting and preserving food and water takes on increasing importance as the length of time in a survival setting increases. With the increased likelihood of diarrhea, replenishing electrolytes becomes critical. For a Marine used to having his provisions issued, foraging can be a significant source of stress.

e. Fatigue. It is essential that survivors employ all available means to preserve mental and physical strength. While food, water, and other energy builders may be in short supply, maximizing sleep is a very controllable factor. Training data collected at MCCDC demonstrates that individuals who lack sleep for 24 hours or more suffer a 25% decrease in effective performance. Further, sleep deprivation directly correlates with increased fear. Effective survival is highly unlikely when fatigue builds to a level where staying awake becomes a stressful evolution.

f. Isolation. Although Marines complain about higher headquarters, we become used to the information and guidance it provides, especially during times of confusion. Being in contact with others provides a greater sense of security and a feeling someone is available to help if problems occur. A significant stressor in survival situations is that often a person or team has to rely solely on its own resources.

4. **NATURAL REACTIONS**. Man has been able to survive many shifts in his environment throughout the centuries. His ability to adapt physically and mentally to a changing world kept him alive. The average person will have some psychological reactions in a survival situation. These are some of the major internal reactions you might experience with the survival stressors.

   a. Fear. Fear is our emotional response to dangerous circumstances that we believe have the potential to cause death, injury, or illness. Fear can have a positive function if it encourages us to be cautious in situations where recklessness could result in injury. Unfortunately, fear can also immobilize a person. It can cause us to become so frightened that we fail to perform activities essential for survival.

   b. Anxiety. Anxiety can be an uneasy, apprehensive feeling we get when faced with dangerous situations. When used in a healthy way, anxiety urges us to act to end, or at least master, the dangers that threaten our existence. A survivor reduces his anxiety by performing those tasks that will ensure his coming through the ordeal alive. Anxiety can overwhelm a Marine to the point where he becomes easily confused and has difficulty thinking.

   c. Anger and Frustration. Frustration arises when a person is continually thwarted in his attempts to reach a goal. The goal of survival is to stay alive until you can reach help or until help can reach you. To achieve this goal, Marines must complete some tasks with minimal resources. One outgrowth of frustration is anger. Getting lost, damaged or forgotten equipment, the weather, inhospitable terrain, enemy patrols, and physical limitations are just a few sources of frustration and anger. Frustration and anger encourage impulsive reactions, irrational behavior, poorly thought-out decisions and, in some instances, an "I quit" attitude. If the Marine does not properly focus his angry feelings, he can waste much energy in activities that do little to further either his chances of survival or the chances of those around him.

d. Depression. Depression is closely linked with frustration and anger when faced with the privations of survival. A destructive cycle between anger and frustration continues until the person becomes worn down-physically, emotionally, and mentally. When a person reaches this point, he starts to give up, and his focus shifts from "What can I do" to "There is nothing I can do." If you allow yourself to sink into a depressed state, then it can sap all your energy and, more important, your will to survive.

e. Loneliness and Boredom. Man is a social animal and enjoys the company of others. Loneliness and boredom can be another source of depression. Marines must find ways to keep their minds productively occupied.

f. Guilt. The circumstances leading to your survival setting are sometimes dramatic and tragic. It may be the result of an accident or military mission where there was a loss of life. Perhaps you were the only, or one of a few, survivors. While naturally relieved to be alive, you simultaneously may be mourning the deaths of others who were less fortunate. Do not let guilt feelings prevent you from living.

5. **PRIORITIES OF WORK IN A SURVIVAL SITUATION.** (FMST.07.19b) Each survival situation will have unique aspects that alter the order in which tasks need to be accomplished. A general guideline is to think in blocks of time.

a. First 24 hours. The first 24 hours are critical in a survival situation. You must make an initial estimate of the situation. Enemy, weather, terrain, time of day, and available resources will determine which tasks need to be accomplished first. They should be the following:

(1) Shelter.

(2) Fire.

(3) Water.

(4) Signaling.

b. Second 24 hours. After the first 24 hours have passed, you will now know you can survive. This time period needs to be spent on expanding your knowledge of the area. By completing the following tasks, you will be able to gain valuable knowledge.

(1) Tools and weapons. By traveling a short distance from your shelter to locate the necessary resources, you will notice edible food sources and game trails.

(2) Traps and snares. Moving further away from your shelter to employ traps and snares, you will be able to locate your shelter area from various vantage points. This will enable you to identify likely avenues of approach into your shelter area.

(3) Pathguards. Knowing the likely avenues of approaches, you can effectively place noise and casualty producing path guards to ensure the security of your shelter area.

c. Remainder of your survival situation. This time is spent on continuously improving your survival situation until your rescue.

6. **GROUP SURVIVAL.** In-group survival, the group's survival depends largely on its ability to organize activity. An emergency situation does not bring people together for a common goal; rather, the more difficult and disordered the situation, the greater are the disorganized group's problems.

a. Groups Morale. High morale must come from internal cohesiveness and not merely through external pressure. The moods and attitudes can become wildly contagious. Conscious, well-planned organization and leadership on the basis of delegated or shared responsibility often can prevent panic. High group morale has many advantages.

(1) Individual feels strengthened and protected since he realizes that his survival depends on others whom he trusts.

(2) The group can meet failure with greater persistency.

(3) The group can formulate goals to help each other face the future.

b. Factors that Influence Group Survival. There are numerous factors that will influence whether a group can successfully survive.

(1) Organization of Manpower - Organized action is important to keep all members of the group informed; this way the members of the group will know what to do and when to do it, both under ordinary circumstances and in emergencies.

(2) Selective Use of Personnel - In well-organized groups, the person often does the job that most closely fits his personal qualifications.

(3) Acceptance of Suggestion and Criticisms - The senior man must accept responsibility for the final decision, but must be able to take suggestion and criticisms from others.

(4) Consideration of Time - On-the-spot decisions that must be acted upon immediately usually determines survival success.

(5) Check Equipment - Failure to check equipment can result in failure to survive.

(6) Acquiring survival knowledge and skills increases survival Knowledge and Skills - Confidence in one's ability to survive.

**UNITED STATES MARINE CORPS**
Mountain Warfare Training Center
Bridgeport, California 93517-5001

FMST.07.23
10/25/01

**STUDENT HANDOUT**

**SIGNALING AND RECOVERY**

**TERMINAL LEARNING OBJECTIVE**. In a cold weather mountainous environment, conduct survival recovery, in accordance with the references. (FMST.07.23)

**ENABLING LEARNING OBJECTIVE.**

(1) Without the aid of reference, select from a given list the audio international distress signal, in accordance with the references. (FMST.07.23a)

(2) Without the aid of references, select from a given list the visual international distress signal, in accordance with the references. (FMST.07.23b)

(3) Without the aid of references, and given a list of messages, match in writing the international code symbols to each corresponding message, in accordance with the references. (FMST.07.23c)

**OUTLINE.**

1. **SIGNALING DEVICES**. The equipment listed below is items that may be on your body or items inside an aircraft. Generally, these items are used as signaling devices while on the move. They must be accessible for use at any moment's notice. Additionally, in a winter mountainous environment, Marines may experience areas that are snow covered and must be familiar with the effects that snow will have on specific signaling devices.

   a. Pyrotechnics. Pyrotechnics include star clusters and smoke grenades. When using smoke grenades in snow pack, a platform must be built. Without a platform, the smoke grenade will sink into the snow pack and the snow will absorb all smoke. A rocket parachute flare or hand flares have been sighted as far away as 35 miles, with an average of 10 miles. Pyrotechnic flares are effective at night, but during daylight their detectability ranges are reduced by 90 percent.

b. M-186 Pen Flare. The M-186 Pen Flare is a signaling device carried in the vest of all crew chiefs and pilots. Remember to cock the gun prior to screwing in the flare.

c. Strobe Light. A strobe light is generally carried in the flight vests of all crew chiefs and pilots. It can be used at night for signaling. Care must be taken because a pilot using goggles may not be able to distinguish a flashing strobe from hostile fire.

d. Flashlight. By using flashlights, a Morse code message can be sent. An SOS distress call consists of sending three dots, three dashes, and three dots. Keep repeating this signal.

e. Whistle. The whistle is used in conjunction with the audio international distress signal. It is used to communicate with forces on the ground.

f. AN/PRC-90 & AN/PRC-112. The AN/ PRC 90 survival radio is a part of the aviator's survival vest. The AN/PRC-112 will eventually replace the AN/PRC-90. Both radios can transmit either tone (beacon) or voice. Frequency for both is 282.8 for voice, and 243.0 for beacon. Both of these frequencies are on the UHF Band.

g. Day/Night Flare. The day/night flare is a good peacetime survival signal. The flare is for night signaling while the smoke is for day. A red cap with three nubbins identifies the older version flare while the new generation has three rings around the body for identification during darkness. The flare burns for approximately 20 seconds while the smoke burns for approximately 60 seconds.

**NOTE**: Once one end is used up, douse in water to cool and save the other end for future use.

h. Signal Mirror. A mirror or any shiny object can be used as a signaling device. It can be used as many times as needed. Mirror signals have been detected as far away as 45 miles and from as high as 16,000', although the average detection distance is 5 miles. It can be concentrated in one area, making it secure from enemy observation. A mirror is the best signaling device for a survivor; however, it is only as effective as its user. Learn how to use one now, before you find yourself in a survival situation.

(1) Military signal mirrors have instructions on the back showing how to use it. It should be kept covered to prevent accidental flashing that may be seen by the enemy.

(2) Any shiny metallic object can be substituted for a signal mirror.

(3) Haze, ground fog, or a mirage may make it hard for a pilot to spot signals from a flashing object. So, if possible, get to the highest point in your area when flashing. If you can't determine the aircraft's location, flash your signal in the direction of the aircraft noise.

AIMING THE SIGNAL MIRROR

2. **METHODS OF COMMUNICATION**

a. Audio. Signaling by means of sound may be good, but it does have some disadvantages.

(1) It has limited range unless you use a device that will significantly project the sound.

(2) It may be hard to pinpoint one's location due to echoes or wind.

(3) International Distress Signal. The survivor will make six blasts in one minute, returned by three blasts in one minute by the rescuer. (FMST.07.23a)

b. Visual. Visual signals are generally better than audio signals. They will pinpoint your location and can been seen at a greater distances under good weather conditions.

(1) The visual international distress symbol is recognized by a series of three evenly spaced improvised signaling devices. (FMST.07.23b)

3. **IMPROVISED SIGNALING DEVICES**. Improvised signaling devices are generally static in nature. They must be placed in a position to be seen by rescuers. They are made from any resources available, whether natural or man-made.

a. Smoke Generator. The smoke generator is an excellent improvised visual signaling device. It gives the survivor the flexibility to signal in either day or night conditions. This type of signal has been sighted as far away as 12 miles, with an average distance of 8 miles. Smoke signals are most effective in calm wind conditions or open terrain, but effectiveness is reduced with wind speeds above 10 knots. Build them as soon as time and the situation permits, and protect them until needed.

(1) Construct your fire in a natural clearing or along the edge of streams (or make a clearing). Signal fires under dense foliage will not be seen from the air.

(2) Find two logs, 6 - 10 inches in diameter, and approximately five feet long. Place the two logs parallel to each other with 3 - 4 feet spacing.

(3) Gather enough sticks, approximately two inches in diameter and four feet long, to lay across the first two logs. This serves as a platform for the fire.

(4) Gather enough completely dry branches to build a pyramid fire. The pyramid fire should be 4 feet by 4 feet by 2 feet high.

(5) Place your tinder under the platform.

(6) Gather enough pine boughs to lay on top of the pyramid fire. This serves to protect the fire and the tinder.

(7) To light, remove the pine bough and ignite the tinder. If available, construct a torch to speed up the lighting process, especially for multiple fires.

SMOKE GENERATOR

(8) To create a smoke effect during the day light hours, place the pine bough on the ignited fire.

(9) Placing a smoke grenade or colored flare under the platform will change the color of the smoke generated. Remember, you want the fire to draw in the colored smoke which will create a smoke color that contrasts with the back ground will increase the chances of success.

b. Arrangement or alteration of natural materials. Such things as twigs or branches can be tramped into letters or symbols in the snow and filled in with contrasting materials. To attract more attention, ground signals should be arranged in big geometric patterns.

(1) International symbols. (FMST.07.23c) The following symbols are internationally known.

| Number | Message | Code symbol |
|---|---|---|
| 1 | REQUIRE ASSISTANCE | V |
| 2 | REQUIRE MEDICAL ASSISTANCE | X |
| 3 | NO OR NEGATIVE | N |
| 4 | YES OR AFFIRMATIVE | Y |
| 5 | PROCEED IN THIS DIRECTION | ↑ |

INTERNATIONAL SYMBOLS

(a) Shadows. If no other means are available, you may have to construct mounds that will use the sun to cast shadows. These mounds should be constructed in one of the International Distress Patterns. To be effective, these shadow signals must be oriented to the sun to produce the best shadows. In areas close to the equator, a North—South line gives a shadow anytime except noon. Areas further north or south of the equator require the use of East—West line or some point of the compass in between to give the best result.

(b) Size. The letters should be large as possible for a pilot or crew to spot. Use the diagram below to incorporate the size to ratio for all letter symbols.

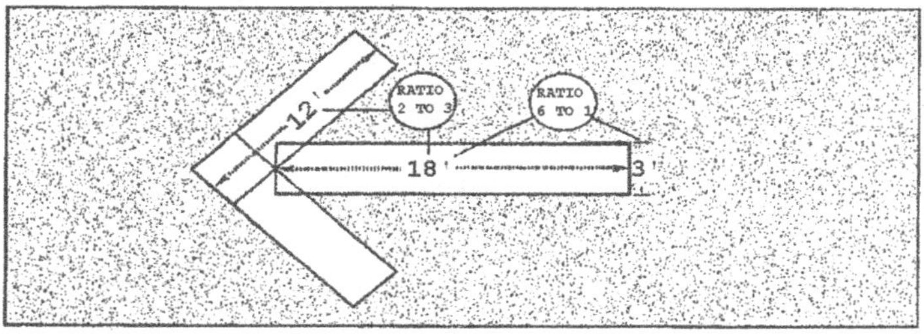

SIZE AND RATIO

(c) Contrast. When constructing letter symbols, contrast the letter from the surrounding vegetation and terrain. Ideally, bring material from another location to build the letter. This could be clothing, air panels, space blanket, etc.

1. On snow, pile bough or use sea dye from a LPP (Life preserver, personal). Fluorescent sea dye markers have been sighted as far away as 10 miles, although the average detection distance is 3 miles.

4. **AIR TO GROUND COMMUNICATIONS**. Air to ground communications can be accomplished by standard aircraft acknowledgments.

a. Aircraft will indicate that ground signals have been seen and understood by:

(1) Rocking wings from side to side. This can be done during the day or in bright moonlight.

b. Aircraft will indicate that ground signals have been seen but not understood by:

(1) Making a complete, clockwise circle during the day or in bright moonlight.

5. **RECOVERY**. Marines trapped behind enemy lines in future conflicts may not experience quick recovery. Marines may have to move to a place that minimizes risk to the recovery force. No matter what signaling device a Marine uses, he must take responsibility for minimizing the recovery force's safety.

a. Placement Considerations. Improvised signaling devices, in a hostile situation, should not be placed near the following areas due to the possibility of compromise:

(1) Obstacles and barriers

(2) Roads and trails.

(3) Inhabited areas.

(4) Waterways and bridges.

(5) Natural lines of drift.

(6) Man-made structures.

(7) All civilian and military personnel.

b. Tactical Consideration. The following tactical considerations should be adhered to prior to employing an improvised signaling device.

(1) Use the signals in a manner that will not jeopardize the safety of the recovery force or you.

(2) Locate a position, which affords observation of the signaling device and facilitates concealed avenues of escape (if detected by enemy forces). Position should be located relatively close to extract site in order to minimize "time spent on ground" by the recovery force.

(3) Maintain continuous security through visual scanning and listening while signaling devices are employed. If weapon systems are available, signaling devices should be covered by fire and/or observation.

(4) If enemy movement is detected in the area, attempt to recover the signaling device, if possible.

(5) Employ improvised signaling devices only during the prescribed times, if briefed in the mission order.

c. Recovery Devices. In mountainous terrain, a helicopter landing may be impossible due to ground slope, snow pack, or vegetation. The survivor must be familiar with recovery devices that may be aboard the aircraft.

JUNGLE PENETRATOR

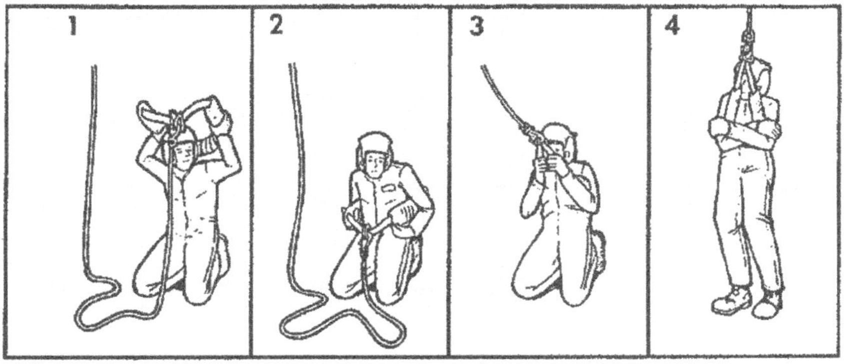


SLING HOIST

d. <u>Recovery by other than aircraft</u>. Recovery by means other than aircraft may occur. Unit SOP's should include signaling and link-up with forces at the following locations:

(1) Border Crossings. The evader who crosses into a neutral country is subject to detention by that country for the duration of the war.

(2) FEBA/FLOT.

(a) Static. Recovery along a static FEBA is always difficult. Under these conditions, enemy and friendly forces can be expected to be densely arrayed and well camouflaged, with good fields of fire. Attempts to penetrate the FEBA should be avoided.

(b) Advancing. Individuals isolated in front of advancing friendly units should immediately take cover and wait for the friendly units to overrun their position.

(c) Retreating. Individuals between opposing forces should immediately take cover and wait for enemy units to pass over their position. After most enemy units have moved on, evaders should try to link up with other isolated friendly elements and return to friendly forces.

(d) Link-up with friendly patrols. Unit authentication numbers and/or locally developed codes may assist the evader to safely make contact in or around the FEBA and when approached by friendly forces.

**UNITED STATES MARINE CORPS**
Mountain Warfare Training Center
Bridgeport, California 93517-5001

FMST.07.04
FMST.07.21
10/25/01

**STUDENT HANDOUT**

# EXPEDIENT SHELTERS AND FIRES

**TERMINAL LEARNING OBJECTIVE**. In a cold weather mountainous environment, construct survival shelters, in accordance with the references. (FMST.07.21)

**TERMINAL LEARNING OBJECTIVE**. Given a survival situation in a wilderness/mountainous environment, and necessary equipment and supplies build a survival/signal fire to meet mission requirements per the references. (FMST.07.04)

**ENABLING LEARNING OBJECTIVES**

(1) Without the aid of references, list in writing the characteristics of a safe survival shelter, in accordance with the references. (FMST.07.21a)

(2) Without the aid of references, list in writing the hazards to avoid when using a natural shelter, in accordance with the references. (FMST.07.21b)

(3) Without the aid of references, list in writing the types of man-made survival shelters, in accordance with the references. (FMST.07.21c)

(4) Without the aid of references, list in writing the tactical fire lay, in accordance with the references. (FMST.07.04a)

(5) In a summer mountainous environment, start a fire using a primitive method, in accordance with the references. (FMST.07.04b)

**OUTLINE.**

**BASIC CRITERIA FOR SHELTER**. (FMST.07.21a) Any type of shelter, whether it is a permanent building, tent age, or a survival shelter, must meet six basic criteria to be safe and effective. The criteria are:

32

a. Protection from the Elements. The shelter must provide protection from rain, snow, wind, sun, etc.

b. Heat Retention. It must have some type of insulation to retain heat; thus preventing the waste of fuel.

c. Ventilation. Ventilation must be constructed, especially if burning fuel for heat. This prevents the accumulation of carbon monoxide. Ventilation is also needed for carbon dioxide given off when breathing.

d. Drying Facility. A drying facility must be constructed to dry wet clothes.

e. Free from Natural Hazards. Shelters should not be built in areas of avalanche hazards, under rock fall or "standing dead" trees have the potential to fall on your shelter.

f. Stable. Shelters must be constructed to withstand the pressures exerted by severe weather.

1. **NATURAL SHELTERS**. Natural shelters are usually the preferred type because they take less time and materials construct. The following may be made into natural shelters with some modification.

   a. Caves or Rock Overhangs. Can be modified by laying walls of rocks, logs or branches across the open sides.

   b. Hollow Logs. Can be cleaned or dug out, then enhanced with ponchos, tarps or parachutes hung across the openings.

   Hazards of Natural Shelters. (FMST.07.21b)

   (1) Animals. Natural shelters may already be inhabited (i.e. bears, coyotes, lions, rats, snakes, etc.). Other concerns from animals may be disease from scat or decaying carcasses.

   (2) Lack of Ventilation. Natural shelters may not have adequate ventilation. Fires may be built inside for heating or cooking but may be uncomfortable or even dangerous because of the smoke build up.

   (3) Gas Pockets. Many caves in a mountainous region may have natural gas pockets in them.

   (4) Instability. Natural shelters may appear stable, but in reality may be a trap waiting to collapse.

2. **MAN-MADE SHELTERS**. (FMST.07.21c) Many configurations of man-made shelters may be used. Limited by imagination and materials available, the following man-made shelters can be used in a survival situation.

a. Snow Wall.

b. Snow Cave.

c. Tree-pit Snow Shelter.

d. Snow Trench.

e. A-frame Shelter.

f. Fallen Tree Bivouac.

g. Snow Coffin.

3. **CONSTRUCTION OF MAN-MADE SHELTERS**. To maximize the shelters effectiveness, Marines should take into consideration the following prior to construction.

a. Considerations

(1) Group Size.

(2) Low Silhouette and reduced living area dimensions for improved heat conservation.

(3) Avoid exposed hilltops, valley floors, moist ground, and avalanche paths.

(4) Create a thermal shelter by applying snow, if available, to roof and sides of shelter.

(5) Location of site to fire wood, water, and signaling, if necessary.

(6) How much time and effort needed to build the shelter.

(7) Can the shelter adequately protect you from the elements (sun, wind, rain, and snow)? Plan on worst-case scenario.

(8) If you have the tools to build it. If not, can you make improvised tools?

(9) If you have the type and amount of materials needed to build it.

(10) When in a tactical environment, you must consider the following:

(a) Provide concealment from enemy observation.

(b) Maintain camouflaged escape routes.

b. Snow Wall. The snow wall is an extremely expedient shelter for one or two men. This shelter is constructed when the elements will not afford time to construct a better shelter.

(1) Basic principles for construction.

(a) Determine wind direction.

(b) Construct a wall of compacted snow in the shape of a horseshoe to shield you from the wind. The wall should be at least 3 feet high and as long as your body.

(c) A poncho or tarp can be attached to the top of the wall with the other end secured to the ground for added protection. Skis, poles, branches, and equipment can be used for added stability.

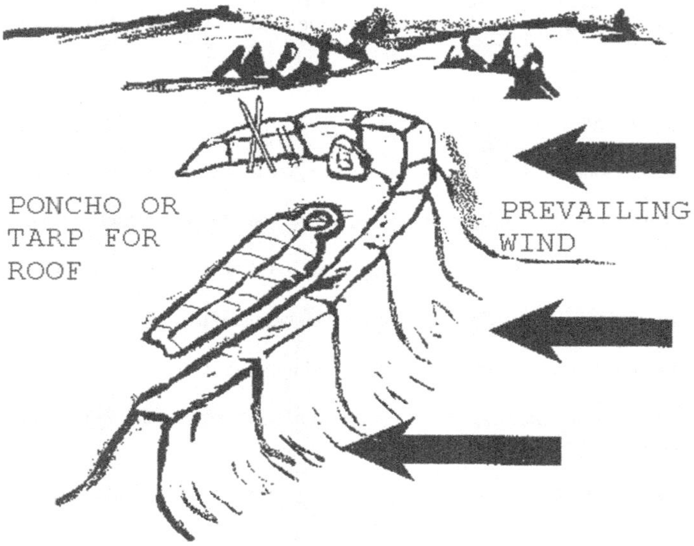


SNOW WALL

c. Snow Cave. A snow cave is used to shelter 1-16 men for extended periods of time. There must be a well-compacted snow pack of at least 6 feet to construct it.

(1) Basic principles for construction and safety precautions.

(a) Dig down into the snow until the desired tunnel entrance has been reached. Place all excavated snow on top of the shelter for added strength.

(b) Cut an entrance opening into the snow approximately 3 feet by 3 feet.

(c) Continue to dig out cave while removing excess snow out of the entrance. Shape the roof into a dome. If a bluish color appears through the snow in the roof, stops, there is not enough snow to support the roof.

(d) Create a cooking/working shelf and a sleeping bench inside the shelter.

(e) A ventilation hole should be dug through the roof at a 45-degree angle above the entrance. A ski pole or branch is left in the hole to mark the hole and allow clearing should the ventilation hole become clogged. A pine bough branch can be placed into the outside of the roof above the hole to aid in keeping the hole clear during falling snow.

(f) Personnel who are digging will become wet from perspiration while digging inside the cave. Personnel that are digging should wear a minimum amount of clothing with a protective layer.

(g) At night, a candle will remain burning while individuals sleep. This will ensure that a proper amount of oxygen exists in the shelter. A fire watch will be assigned to check each cave's candle. If a candle is found distinguished, the occupants of the shelter will be awakened to ensure their safety. The candle will then be re-lit.

(h) A shovel will be located immediately inside and outside the snow cave's opening. This is for quick extraction of snow in case of a collapsed cave or accumulated snowfall.

(i) Packs, poncho, or snow blocks can be used to block the entrance to the cave.

(j) Crossed skis or snowshoes will mark entrance to the cave.

(k) Occupants should sleep with their head towards the opening.

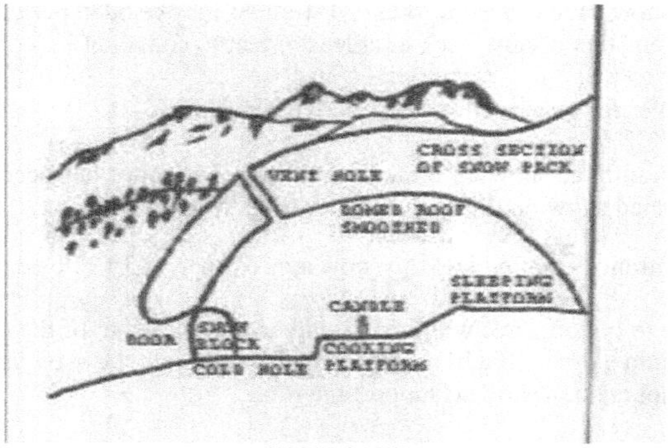

SNOW CAVE

d. Tree-pit Snow Shelter. A tree-pit snow shelter is designed for 1-3 men for short periods of time. It provides excellent overhead cover and concealment and should be used for Listening Post/Ops.

(1) Basic principles for construction.

(a) Locate a tree with bushy branches that provides overhead cover.

(b) Dig out the snow around the tree trunk until you reach the depth and diameter desired, or until you reach the ground.

(c) Find and cut other evergreen boughs. Place them over the top of the pit for additional concealment.

(d) Place evergreen boughs in the bottom of the pit for insulation.

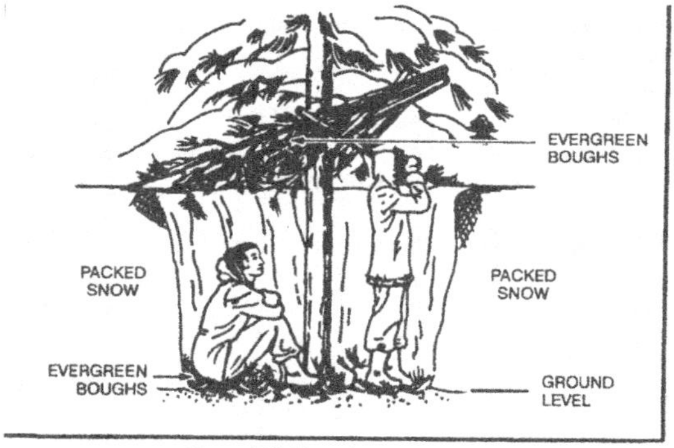


TREE-PIT SHELTER

e. Fallen Tree Bivouac. The fallen tree bivouac is an excellent shelter because most of the work has already been done.

(1) Ensure the tree is stable prior to constructing.

(2) Branches on the underside are cut away making a hollow underneath.

(3) Place additional insulating material to the top and sides of the tree.

(4) A small fire is built outside of the shelter.

FALLEN TREE BIVOUAC

f. A-Frame Shelter. An A-Frame shelter is constructed for 1-3 individuals. After the framework is constructed, pine bough/tent age is interwoven onto the frame and snow is packed onto the outside for insulation.

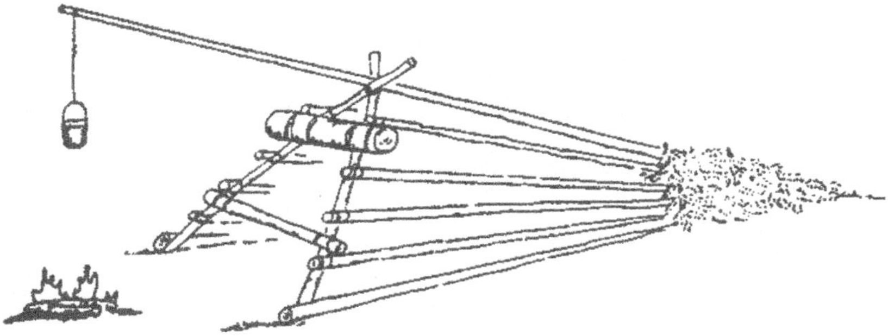

A-FRAME

g. Snow Trench. A snow trench is a short-term shelter used on extremely hard pack snow and when trees or building materials are not available, (i.e., Alpine and Glaciers). Blocks of snow or ice are cut and placed to build this shelter

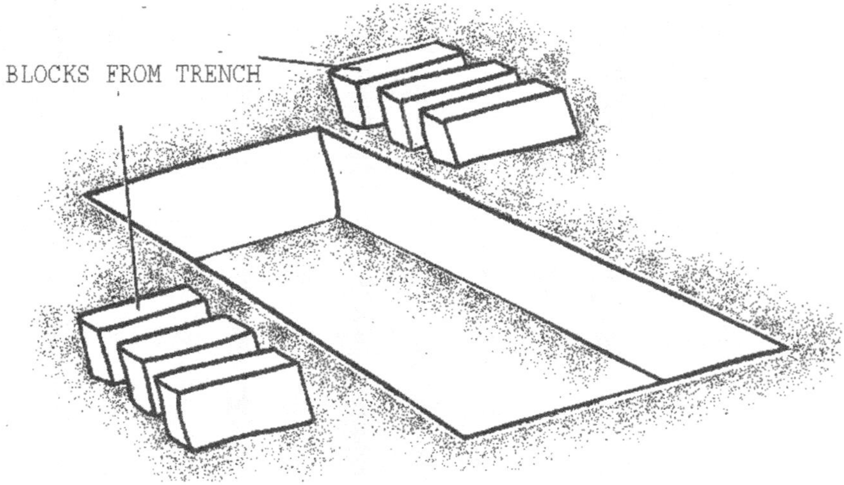


SNOW TRENCH (FIRST STEP)

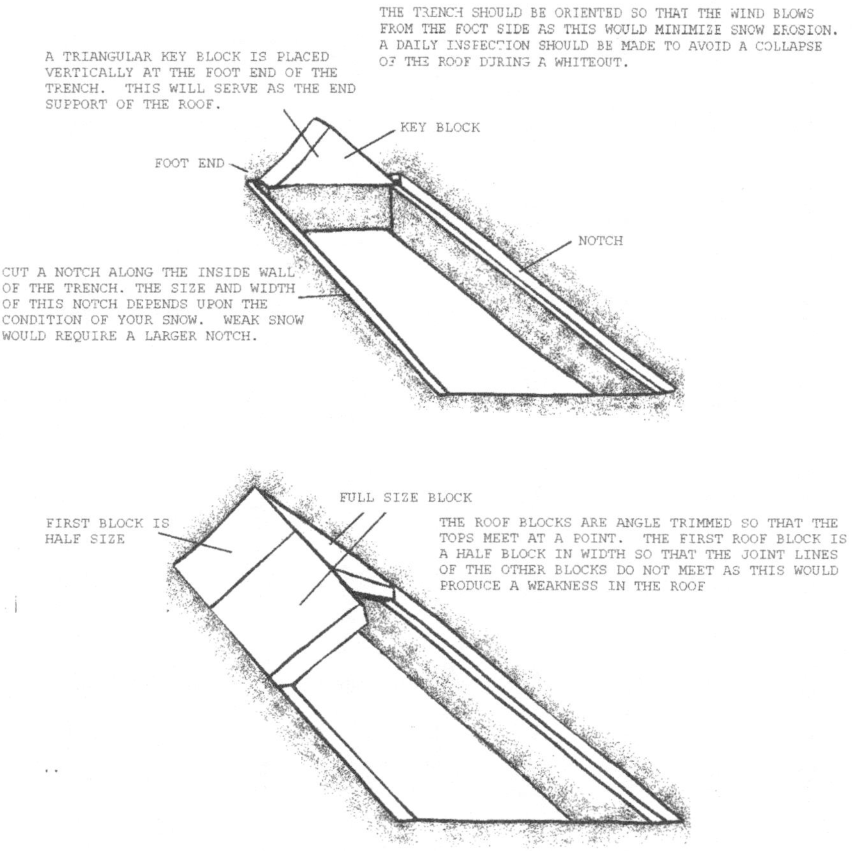


SNOW TRENCH

h. Snow Coffin. A snow coffin is built for 1-4 men for extended periods of time. It is a variation of the snow trench and A-frame, which requires at least 4 feet of compacted snow.

(1) Basic principles for construction.

(a) Dig a trench into the snow approximately 3 feet wide, 8-12 feet long, and 4 feet deep.

(b) Dig a cold hole into the floor of the trench and sleeping platforms (coffins) off the sides of the trench.

(c) Cover the top of the trench for added protection with either an A-frame or poncho/tarp.

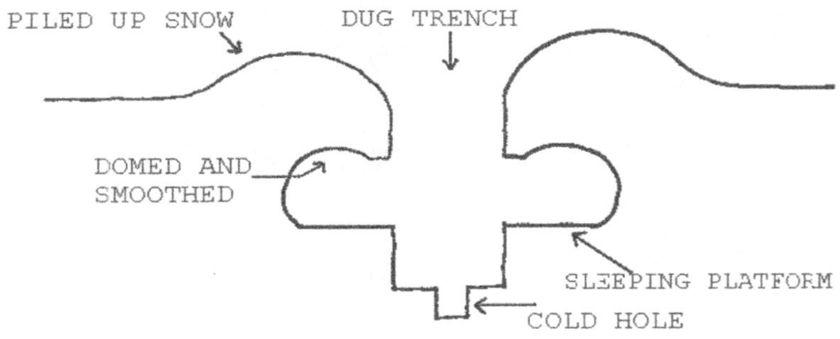


SNOW COFFIN

i. The following table can be used as a general guideline to determine which shelter to construct.

| SNOW PACK | SNOW DEPTH | EST. HRS. TO CONSTRUCT | RECOMMENDED SHELTER |
|---|---|---|---|
| LOOSE | < 2 FEET | 2 | A-FRAME |
| COMPACTED | 4-6 FEET | 2-3 | SNOW COFFIN |
| COMPACTED | > 6 FEET | 3 | SNOW CAVE |
| ICED | N/A | 2-3 | SNOW TRENCH |
| N/A | N/A | 1-2 | FALLEN TREE |
| N/A | > 4 FEET | 1-2 | TREE-PIT |
| N/A | > 2 FEET | 30 MIN | SNOW WALL |

**NOTE**: The Fallen Tree Bivouac and Tree-pit Shelter can only be used if a suitable site has been located.

4. **SURVIVAL FIRES**. Fires fall into two main categories: those built for cooking and those built for warmth and signaling. The basic steps are the same for both: preparing the fire lay, gathering fuel, building the fire, and properly extinguishing the fire.

a. Preparing the fire lay. There are two types of fire lays: fire pit and Dakota hole. Fire pits are probably the most common.

(1) Without a platform in the snow, the fire will sink. Create a platform by:

(a) Lay several green logs side by side for the size of your fire.

(b) Place a couple of inches of snow over the top of the green logs.

(c) Place a second row of green logs, side-by-side, perpendicular to the first row.

(d) Build your fire on top of the platform.

(2) Create a windbreak to confine the heat and prevent the wind from scattering sparks. Place rocks or logs used in constructing the fire lay parallel to the wind. The prevailing downwind end should be narrower to create a chimney effect.

(3) Avoid using wet rocks. Heat acting on the dampness in sandstone, shale, and stones from streams may cause them to explode.

(4) Dakota Hole. (FMST.07.04a) The Dakota Hole is a tactical fire lay. Although no fire is 100% tactical, this fire lay will accomplish certain things:

(a) Reduces the signature of the fire by placing it below ground.

(b) Provides more of a concentrated heat source to boil and cook, thus preserving fuel and lessening the amount of burning time.

(c) By creating a large air draft, the fire will burn with less smoke than the fire pit.

(d) It is easier to light in high winds.

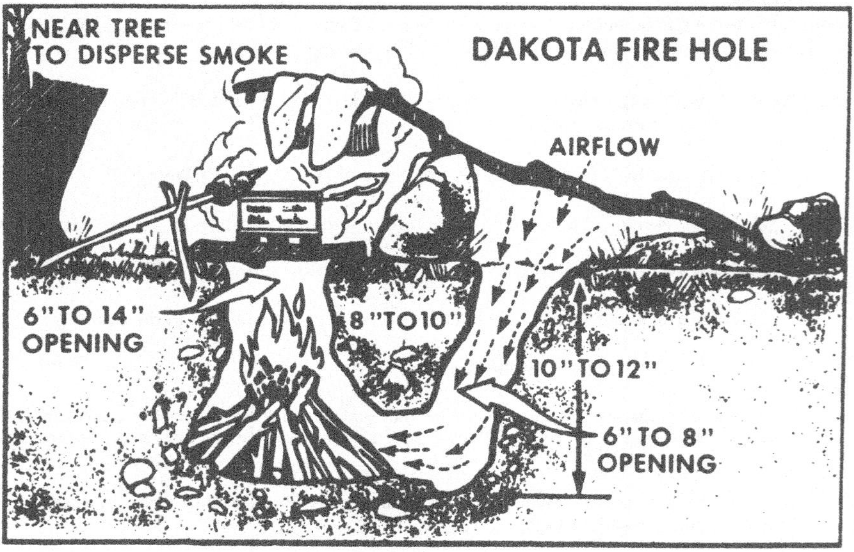


DAKOTA HOLE

b. Gather Fuel. Many Marines take shortcuts when gathering firewood. Taking a few extra minutes can mean the difference between ease and frustration when building a fire.

(1) Tinder. Tinder is the initial fuel. It should be fine and dry. Gather a double handful of tinder for the fire to be built and an extra double handful to be stored in a dry place for the following morning. Dew can moisten tinder enough to make lighting the fire difficult. Some examples are:

(a) Shredded cedar/juniper bark, pine needles.

(b) Dry grass.

(c) Slivers shaved from a dry stick.

(d) Hornet's nest.

(e) Natural fibers from equipment supplemented with pine pitch (i.e., cotton battle dressing).

(f) Cotton balls and petroleum jelly or Char-cloth.

**NOTE**: Sticks used for tinder should be dry and not larger than the diameter of a toothpick.

(2) Kindling. This is the material that is ignited by the tinder that will burn long enough to ignite the fuel.

(a) Small sticks/twigs pencil-thick up to the thickness of the thumb. Ensure that they are dry.

(b) Due to a typically large resin content, evergreen limbs often make the best kindling. They burn hot and fast, but typically do not last long.

(3) Fuel Wood. Fuel Wood is used to keep the blaze going long enough to fulfill its purpose. Ideally, it should burn slow enough to conserve the woodpile, make plenty of heat, and leave an ample supply of long-lasting coals.

(a) Firewood broken from the dead limbs of standing trees or windfalls held off the ground will have absorbed less moisture and therefore should burn easily.

(b) Refrain from cutting down live, green trees.

(c) Softwoods (evergreens and conifers) will burn hot and fast with lots of smoke and spark, leaving little in the way of coals. Hardwoods (broad leaf trees) will burn slower with less smoke and leave a good bed of coals.

(d) Learn the woods indigenous to the area. Birch, dogwood, and maple are excellent fuels. Osage orange, ironwood, and manzanita, though difficult to break up, make terrific coals. Aspen and cottonwood burn clean but leave little coals.

(e) Stack your wood supply close enough to be handy, but far enough from the flames to be safe. Protect your supply from additional precipitation.

(f) If you happen to go down in an aircraft that has not burned, a mixture of gas and oil may be used. Use caution when igniting this mixture.

c. Building the Fire. The type of fire built will be dependent upon its intended use either cooking or heating and signaling.

(1) Cooking Fires. The following listed fires are best used for cooking:

(a) Teepee Fire. The teepee fire is used to produce a concentrated heat source, primarily for cooking. Once a good supply of coals can be seen, collapse the teepee and push embers into a compact bed.

TEEPEE FIRE

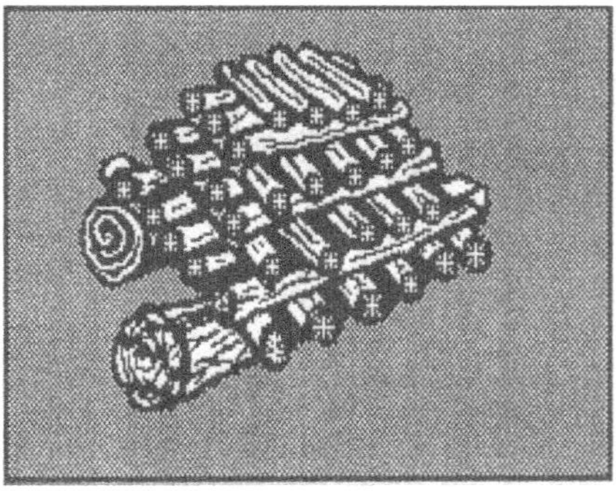

PYRAMID FIRE

(2) Heating Fires.

(a) Pyramid Fire. Pyramid fires are used to produce large amounts of light and heat, to dry out wet wood, and provide coals for cooking.

(3) Starting Fires. Lighting fires falls into two categories modern methods and primitive methods.

(a) Modern Methods. Modern igniters use modern devices we normally think of to start a fire. Reliance upon these methods may result in failure during a survival situation. These items may fail when required to serve their purpose.

1. Matches and Lighters. Ensure you waterproof these items.

2. Convex Lens. Binocular, camera, telescopic sights, or magnifying lens are used on bright, sunny days to ignite tender.

3. Flint and Steel. Sometimes known as metal matches or "Mag Block". Scrape your knife or carbon steel against the flint to produce a spark onto the tinder. Some types of flint & steel designs will have a block of magnesium attached to the device, which can be shaved onto the tinder prior to igniting. Other designs may have magnesium mixed into the flint to produce a higher quality of spark.

(b) Primitive Methods. Primitive fire methods are those developed by early man. There are numerous techniques that fall into this category. The only method that will be taught at MCMWTC is the Bow & Drill.

1. Bow & Drill. The technique of starting a fire with a bow & drill is a true field expedient fire starting method, which requires a piece of cord and knife from

your survival kit to construct. The components of the bow & drill are bow, drill, socket, fireboard, ember patch, and bird's nest.

- Bow. The bow is a resilient, green stick about 3/4 of an inch in diameter and 30-36 inches in length. The bowstring can be any type of cord, however, 550 cord works best. Tie the string from one end of the bow to the other, without any slack.

- Drill. The drill should be a straight, seasoned hardwood stick about 1/2 to 3/4 of an inch in diameter and 8 to 12 inches in length. The top end is tapered to a blunt point to reduce friction generated in the socket. The bottom end is slightly rounded to fit snugly into the depression on the fireboard.

- Socket. The socket is an easily grasped stone or piece of hardwood or bone with a slight depression on one side. Use it to hold the drill in place and to apply downward pressure.

- Fire board. The fire board is a seasoned softwood board which should ideally be 3/4 of an inch thick, 2-4 inches wide, and 8-10 inches long. Cut a depression 3/4 of an inch from the edge on one side of the fireboard. Cut a U-shape notch from the edge of the fireboard into the depression. This notch is designed to collect and form an ember, which will be used to ignite the tinder.

- Ember Patch. The ember patch is made from any type of suitable material (i.e., leather, aluminum foil, bark). It is used to catch and transfer the ember from the fireboard to the bird's nest. Ideally, it should be 4 inches by 4 inches in size.

- Birds Nest. The bird's nest is a double handful of tinder, which will be made into the shape of a nest. Tinder must be dry and finely shredded material (i.e., outer bark from juniper/cedar/sage brush or inner bark from cottonwood/aspen or dry grass/moss). Lay your tinder out in two equal rows about 4 inches wide and 8-12 inches long. Loosely roll the first row into a ball and knead the tinder to further break down the fibers. Place this ball perpendicular onto the second row of tinder and wrap. Knead the tinder until all fibers of the ball are interwoven. Insert the drill half way into the ball to form a partial cylinder. This is where the ember will be placed.

2. Producing a fire using the bow & drill.

    - Place the ember patch under the U-shaped notch.

- Assume the kneeling position, with the left foot on the fireboard near the depression.

- Load the bow with the drill. Ensure the drill is between the wood of the bow and bowstring. Place the drill into the depression on the fireboard. Place the socket on the tapered end of the drill.

- Use the left hand to hold the socket while applying downward pressure.

- Use the right hand to grasp the bow. With a smooth sawing motion, move the bow back and forth to twirl the drill.

- Once you have established a smooth motion, smoke will appear. Once smoke appears, apply more downward pressure and saw the bow faster.

- When a thick layer of smoke has accumulated around the depression, stop all movement. Remove the bow, drill, and socket from the fireboard, without moving the fireboard. Carefully remove your left foot off the fireboard.

- Gently tap the fireboard to ensure all of the ember has fallen out of the U-shaped notch and is lying on the ember patch. Remove the fireboard.

- Slowly fan the black powder to solidify it into a glowing ember. Grasping the ember patch, carefully drop the ember into the cylinder of the bird's nest.

- Grasp the bird's nest with the cylinder facing towards you and parallel to the ground. Gently blow air into the cylinder. As smoke from the nest becomes thicker, continue to blow air into the cylinder until fire appears.

3. <u>Trouble Shooting the Bow & Drill</u>

- Drill will not stay in depression- Apply more downward pressure and/or increase width/depth of depression.

- Drills will not twirl- Lessen the amount of downward pressure and/or tighten bowstring.

- Socket smoking- Lessen the amount of downward pressure. Wood too soft when compared to hardness of drill. Add some lubrication: animal fat, oil, or grease.

- No smoke- Drill and fireboard are the same wood. Wood may not be seasoned. Check drill to ensure that it is straight. Keep left hand locked against left shin while sawing.

- Smoke but no ember- U-shaped notch not cut into center of the depression.

- Bowstring runs up and down drill- Use a locked right arm when sawing. Check drill to ensure that it is straight. Ensure bowstring runs over the top of the left boot.

- Birds nest will not ignite- Tinder not dry. Nest woven too tight. Tinder not kneaded enough. Blowing too hard (ember will fracture).

d. Extinguishing the Fire. The fire must be properly extinguished. This is accomplished by using the drown, stir, and feel method.

(1) Drown the fire by pouring at water in the fire lay.

(2) Stir the ember bed to ensure that the fire is completely out.

(3) Check the bed of your fire by feeling for any hot spots.

(4) If any hot spots are found, start the process all over again.

**UNITED STATES MARINE CORPS**
Mountain Warfare Training Center
Bridgeport, California 93517-5001

FMST.07.07
08/15/01

**STUDENT HANDOUT**

# SKIJORING

**TERMINAL LEARNING OBJECTIVE**. Given snow covered mountainous terrain, a BV, 165 ft static rope, military skis and an assault load, conduct skijoring operations, in accordance with the references. (FMST.07.07)

**ENABLING LEARNING OBJECTIVES**.

(1) Without the aid of references, select from a given list the advantages of skijoring, in accordance with the references. (FMST.07.07a)

(2) Without the aid of references, select from a given list the disadvantages of skijoring, in accordance with the references. (FMST.07.07b)

(3) Without the aid of references, list in writing four safety requirements for skijoring, in accordance with the references. (FMST.07.07c)

**OUTLINE.**

1. **ADVANTAGES**. (CWM.6.30a)

a. Each skier expends less energy if skijoring is done properly. It's similar to water skiing behind a boat. However, if you keep falling down and picking yourself up time after time, you'll be defeating the purpose.

b. Movement of a ski borne unit can be expedited by skijoring. An over the snow vehicle is going to travel faster than a unit skiing. However, improperly trained ski borne troops can be the exception.



2. **DISADVANTAGES**. (FMST.07.07b)

a. Skijoring improperly can become tiring and slow. Thus, training is critical and can take a lot of time. Proper training is important not only for the skiers, but also for the drivers. Improperly trained drivers could cause serious accidents to skiers.

b. Cold weather injuries can occur anytime in a cold weather environment. When Skijoring, the wind chill against a skier is going to be increased significantly. Ski troops must be familiar with proper clothing protection to prevent injury.

3. **SAFETY REQUIREMENTS**. (FMST.07.07c) The following safety requirements must be adhered to:

a. Never under any circumstances will a skier tie into a skijoring rope. This could cause the skier to be dragged over, around, or through obstacles and cause serious injuries.

b. The rate of speed for a vehicle pulling experienced skiers should not exceed 25 mph. For inexperienced skiers, the rate of speed should not exceed 15 mph.

c. Skiers should be spaced at least 1/2 ski length apart from the tail of the front skier to the tip of the following skier. This will give the skiers ample time to react to most situations, such as a fallen skier or rounding a corner.

d. A safety rider will always be employed. He should be experienced in skijoring and must have visual contact with skiers and communications with the driver. Anytime a group of Marines are skijoring, a safety rider is required and his tasks include:

(1) Observation. Be situated where he can observe all skiers.

(2) Communication. Be in communication with the driver at all times, whether by using a whistle, a cord, or another man.

(3) Halt signal. Give the halt signal for all others to hear in case anyone falls, and most importantly, so the driver knows that he should halt.

(4) Other signals. Have a signal that everyone knows for stopping, slowing, accelerating, etc.

4. **PREPARING THE OVER THE SNOW VEHICLE**. The over the snow vehicles used at MWTC are the DMC 1450, LMC 1200, SUSV M-973 (BV-206), HUMMV equipped with Mac Track. The preparations we will be discussing will remain the same for these vehicles; the only exception is the load towing capacity, which will be determined by the driver of the vehicle based upon the snow conditions, and also the position of the safety rider.

a. Mirrors. Ensure that the mirrors for the driver are in correct alignment for best viewing to the rear of the vehicle.

b. Back-up Horn. Ensure that the vehicle has a back-up horn and that it is functional.

c. TC Hatch. The safety rider occupies the TC Hatch so he can observe the skiers.

d. Towing Capacity. Due to the different towing capacities of the vehicles used here, the amount of skiers to be the driver will determine skijored. To attach the skijoring rope to any one of these vehicles remain the same as stated below:

(1) Towing rope. Find the middle of the towing rope.

(2) Clove hitch. Open the trailer hitch and place a clove hitch into it using the middle of the towing rope. A clove hitch should be the only knot utilized.

(3) Trailer hitch. Tighten down the clove hitch and replace the top of the trailer hitch and ensure it is locked down to prevent the rope from slipping out.

(4) Figure of eight loops. These should be tied on the end of each line of the towing rope for the last men.

5. **SKIJORING TECHNIQUES**

a. Skiers. Each skier should line up on the outside of his prospected line. The first skiers should be at least 1 ski length behind the vehicle or at a vantage where the safety rider can observe him.

b. Proper Interval. The proper interval of 1/2 ski length apart from each other should be utilized before hooking up.

c. To Hook Up. Once the interval is appropriate and the skiers are ready to hook up, the forward skiers should hook up first, then the second set, etc., until the last skiers hook up.

(1) Ski poles. To do this, place the ski pole handgrips on top of the towrope. Next grasp the towrope in front of the pole and wrap the rope around your ski poles and directly under the handgrips, forming a half hitch. Place the ski pole baskets behind your outside armpit to use as a rest.

(2) Grip the rope. Do this with the inside hand.

d. Keep your skis parallel. They should be about shoulder width apart; knees should be flexed in the basic ski stance.

e. Last skiers. The last skiers should ensure they keep the tension on the end of the rope by hanging on to the figure of eight loops. This will keep their ski poles in place.

f. If someone falls, everyone lets go of the rope and skis outboard.

g. Put experienced skiers in front and back of the rope.

h. Shuffle skis when starting to prevent skis from freezing to the ground.

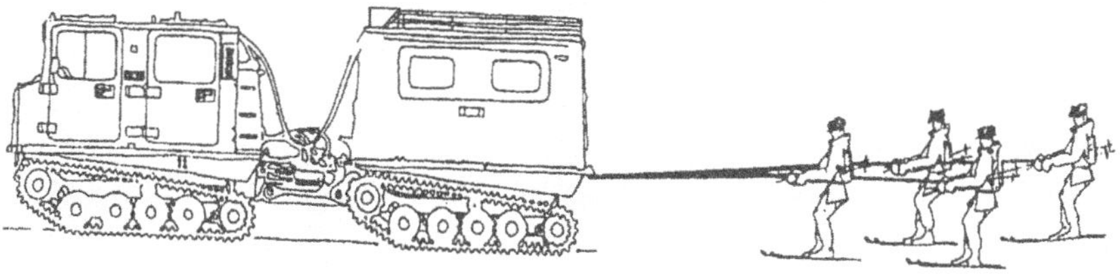

SKIJORING

6. **SPECIAL CONSIDERATIONS**

a. Communication signals between safety rider and driver must be established.

b. The safety rider must be positioned to view all skijoring.

c. Start slowly with a consistent speed and stop gradually. Increase speed down gentle slopes and skiers will unhook before steep slopes. The decision to unhook for downhill grades will depend upon the skier's ability.

d. Sharp turns must be avoided.

e. Visibility might warrant fewer skiers.

f. Snow Conditions. In deeper snow, the vehicle might have to pull fewer skiers.

g. Skier's ability will naturally dictate the speed of the vehicle.

h. If sleds must be towed, it is best if they are towed separately for safety, and a safety rider will be utilized.

i. When negotiating turns or curves, ski to the outside of the turn, especially the rear skiers on the inside rope.

j. Ensure drivers are properly briefed.

Made in the USA
Monee, IL
04 November 2026

67383973R10164